Lung-an Ying

**Infinite
Element
Methods**

Lung-an Ying

Infinite Element Methods

Peking University Press

Vieweg Publishing

Mathematical Subject Classification: 65-XX

All rights reserved
© Peking University Press, Beijing, 1995, and
© Friedr. Vieweg & Sohn Verlagsgesellschaft mbH, Braunschweig/Wiesbaden, 1995

Vieweg is a subsidiary company of the Bertelsmann Professional Information.

No part of this publication may be reproduced, stored in a retrieval system or transmitted, mechanical, photocopying or otherwise, without prior permission of the copyright holder.

Typesetting: Peking University Press
Printing and binding: C & C OFFSET PRINTING CO.. LTD. (H. K.)
Printed in Hong Kong

ISBN 7-301-02781-8/O · 350
ISBN 3-528-06610-5

Preface

As its name indicates, in the infinite element method the underlying domain is divided into infinitely many pieces. This leads to a system of infinitely many equations for infinitely many unknowns; but these can be reduced by analytical techniques to a finite system when some sort of scaling is present in the original problem. The simplest illustrative example, described carefully at the beginning of the first chapter of the book before us, is the solution of the Dirichlet problem in the exterior of some polygon. The exterior is subdivided into annular regions by a sequence of geometrically expanding images of the given polygon; these annuli are then further subdivided. The resulting variational equations take the form of a block tridiagonal Toeplitz matrix, with an inhomogeneous term in the zero component. Various efficient methods are described for solving such systems of equations.

Another class of problems for which the infinite element method is appropriate is the resolution of singularities, such as occur in elastic structures. The theory of elasticity was the starting point of the body of work described in this book; however, a glance at the table of contents reveals that the scope of the method now embraces a very large class of elliptic and parabolic problems.

Chapter 2 is devoted to the precise description of the algorithms used in a wide variety of contexts. Chapter 3 uses subtle analytical techniques to deal with the mathematical aspects of convergence and error estimates. Numerical examples are described in Chapter 4; modest in size by present-day standards, these examples nevertheless domonstrate convincingly the practical power of the method.

The infinite element method is, whereever applicable, an elegant and efficient approach to solving problems in physics and engineering. Professor Ying's welcome book makes it available to the community of numerical analysts and computational scientists.

New York, 1994
Peter D. Lax

Foreword

The infinite element method is a combination of an infinite partition technique and the finite element method. To the author's knowledge the pioneering work of the infinite partition technique was due to Silvester and Cermak [1] in 1969 which was stated in the category of the finite difference method. The first paper on this combination was [2] by Thatcher in 1975. Independently a paper by the author [4] was first presented at the National Conference on Fracture Mechanics in Nanning, China in 1974, then it was published in 1977, and the English version of it appeared in 1978. Afterwards this method has been studied extensively by many authors. For instance, Han Houde has done some joint work with the author on the iterative method and exterior problems, Shao Xiumin and Wang Shujing have introduced the Fourier method, Pan Hao has done some joint work with the author on the application to fracture mechanics, and Xu Jinchao has obtained some interesting results on convergence, and others. Without the joint effort of these researchers, the infinite element method could not be developed to such a scale as it is today.

The author's book on the infinite element method was published in Chinese by the Peking University Press in 1992 which contained the results before 1989. Since then new results have been obtained, and the author is very glad to learn that this book has attracted more young researchers to this method. In this present English version many supplements to the Chinese edition have been included.

The basic idea of the infinite element method is very simple: to solve a problem with the finite element method, one should divide the object into some elements, and restricted by the capacity and the speed of the computer, the number of elements is limited. This restriction brings difficulty in solving some kinds of problems. For example we intend to compute a structure on an infinitely large basis. It is hard to make a partition on the basis. Too few unknowns cause losing accuracy, and too many unknowns exceed the ability of the computer, or cause a lot of waste. Another example is stress concentration. Mechanical quantities vary violently at some local points of the structure. The local refinement is also restricted, which causes inaccuracy in computation. The infinite element method breaks this restriction, which allows using an infinite number of elements if necessary. So it is more flexible. For a given problem if it is sufficient to use the finite element method, then we use the convensional approach, otherwise we can use infinite elements. Different problems are treated differently. Moreover for the solutions with singularities the infinite element solutions possess singularities near the singular points too, and qualitatively and quantitatively these singularities are close to those of the exact solutions.

Thus the core of the infinite element method is to solve an infinite number of elements. There are fruitful results and approaches to deal with "infinity" in mathematics, therefore this difficulty can be overcome if we make use of these results appropriately. The readers will find that after applying some ingenious tricks, to deal with a problem with an infinite number of elements we do not require an "infinitely large computer". On the contrary a personal computer is usually sufficient for solving common problems. That is the attractiveness of the infinite element method.

Of course to treat those problems there are some other methods, such as the boundary element method, the semi-analytic method, and others. Compared to these methods the advantage of the infinite element method is that it does not need the expressions of the exact solutions. It is well known that such expressions are hard to obtain in the most cases for partial differential equations. The calculating of stress intensity factors may be the only exception, the readers will find the expansions of the solutions near singular points are useful to obtain the approximate stress intensity factors.

Related to the properties of the infinite element method the mathematical theory has been established. Applying the results of weighted Sobolev spaces, the author proved that the rate of convergence of the infinite element method for singular solutions is the same as that of the the finite element method for smooth solutions, and some times it is even higher. Applying Kato's perturbation theory for linear operators, the author proved that the infinite element solutions can be expanded near singular points and these expansions converge to those of exact solutions in a term by term manner. This result shows that the infinite element method is a natural approach for solving singularities.

Taking account of the different needs of different readers, the author arranges the material in this book as follows: Chapter One and Chapter Four are especially provided for those readers who are interested in applications only. A detailed statement of the algorithm of the infinite element method is contained in Chapter One, and to read this chapter one only needs the knowledge of calculus and matrix operation. Some examples to show the effect of the infinite element method are collected in Chapter Four. However Chapter Two and Chapter Three are more specialized. The algebraic systems generated by the infinite element method are studied in Chapter Two, and to read this chapter one needs more knowledge of mathematics, for instance the theory of the Sobolev spaces and the matrix theory. Convergence results are proved in Chapter Three, and to read this chapter one needs further knowledge, for instance the theory of the finite element method.

The author wishes to express his sincere gratitude to Professor Duan Xuefu who recommended his first paper on the infinite element method to the journal "Scientia Sinica" during the chaos age of China. His cordial thanks are due to Professor Feng Kang who as a referee read his first paper carefully and helped him to improve it. Afterwards he has got much encouragement from Professor Feng. Professor Feng passed away on August 17, 1993. The author greatly misses him, and it is sad that Professor Feng can not see the English version of this book.

The author expresses his hearty thanks tho Professor Guo Zhongheng. They started to work on fracture mechanics together. The author has learnt a lot about mechanics from Professor Guo. Under Professor Guo's encouragement the author submitted his first paper on the infinite element method for publication. Professor Guo passed away on September 22, 1993. The author also greatly misses him, and always remembers the time they worked together.

The author takes pleasure in acknowledging his friends, Professor Han Houde and Dr. Pan Hao who collaborated with him to develop the infinite element method.

The author appreciates the helpful assistance of his former students Dr. Xu Jinchao, Mr. Lu Jianqun, Dr. Jiang Huaqiong, Mr. Liu Weidong, Mr. Wei Wanming, Mr. Xie Songfeng, Mr. Wu Dongbing, and Dr. Feng Hui who have worked with him on the infinite element method and made various contributions.

Finally the author would like to thank the editor Qiu Shuqing for the excellent quality of this edition.

Beijing, spring, 1994
Lung-an Ying

Contents

1 Algorithm   1
  1.1 Two dimensional exterior problems of the Laplace equation   1
  1.2 Fourier method   6
  1.3 Iterative method   8
  1.4 General elements   11
  1.5 Three dimensional exterior problems of the Laplace equation   13
  1.6 Problems on other unbounded domains   15
  1.7 Corner problems   15
  1.8 Plane elasticity problems   18
  1.9 Calculation of stress intensity factors   26
  1.10 Exterior Stokes problems   30
  1.11 Nonhomogeneous equations and nonhomogeneous boundary conditions   37
  1.12 Boundary value problems of the biharmonic equation   41
  1.13 Multigrid algorithm   44
  1.14 Helmholtz equation   46
  1.15 Elliptic equations with variable coefficients   52
  1.16 Parabolic equations   55
  1.17 Variational inequalities   56

2 Foundations of algorithm   59
  2.1 Infinite element spaces   59
  2.2 Transfer matrices   64
  2.3 Further discussion for the infinite element spaces and the transfer matrices   67
  2.4 Transfer matrices for the plane elasticity problems   71
  2.5 Combined stiffness matrices   74
  2.6 Structure of the general solutions   75
  2.7 Block circular stiffness matrices   81
  2.8 Iterative method of the first type   83
  2.9 Iterative method of the second type   86
  2.10 General elliptic systems   92
  2.11 Exterior Stokes problems (II)   99
  2.12 Nonhomogeneous equations and the Helmholtz equation   106
  2.13 Discontinuous boundary value problems   116
  2.14 Structure of the solutions near a singular point to interface problems   118

3 Convergence — 121
- 3.1 Some auxiliary inequalities — 121
- 3.2 Approximate properties of piecewise polynomials — 124
- 3.3 H^1 and L^2 convergence — 127
- 3.4 Maximum principle and uniform convergence — 131
- 3.5 A superconvergence estimate — 137
- 3.6 Convergence by terms near the singular point — 152
- 3.7 Multigrid algorithm (II) — 162
- 3.8 Parabolic equations (II) — 167
- 3.9 Variational inequalities (II) — 174

4 Examples — 181
- 4.1 A glance at the error — 181
- 4.2 Boundary value problems and eigenvalue problems — 183
- 4.3 Equations with variable coefficients — 186
- 4.4 Stress intensity factors — 188
- 4.5 Stokes external flow — 190
- 4.6 Square driven cavity — 193
- 4.7 Navier-Stokes external flow — 197
- 4.8 A numerical solution to a variational inequality — 202

Bibliography — 205

1 Algorithm

1.1 Two dimensional exterior problems of the Laplace equation

We start with the following model problem.

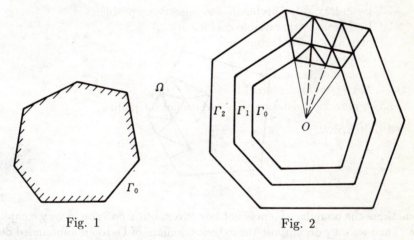

Fig. 1   Fig. 2

Let Γ_0 be a convex polygon in a plane, and domain Ω be its exterior (Fig.1). We solve the Dirichlet problems of the Laplace equation

$$\triangle u = 0, \qquad (x,y) \in \Omega, \qquad (1.1.1)$$

$$u|_{\Gamma_0} = f \qquad (1.1.2)$$

on Ω, or the Neumann problems, where the boundary condition (1.1.2) is replaced by

$$\left.\frac{\partial u}{\partial \nu}\right|_{\Gamma_0} = f, \qquad (1.1.3)$$

Here and hereafter we always denote by ν the outward normal direction. The strain energy associated with (1.1.1) is

$$W = \frac{1}{2} \int_\Omega \left\{ \left(\frac{\partial u}{\partial x}\right)^2 + \left(\frac{\partial u}{\partial y}\right)^2 \right\} dxdy. \qquad (1.1.4)$$

1 ALGORITHM

To solve (1.1.1) (1.1.2) or (1.1.1) (1.1.3), Ω is divided into an infinite number of triangular elements. We may assume the origin is at the interior of Γ_0. Taking a constant $\xi > 1$, we draw the similar curves of Γ_0 with center O and the constants of proportionality $\xi, \xi^2, \cdots, \xi^k, \cdots$, which are denoted by $\Gamma_1, \Gamma_2, \cdots, \Gamma_k, \cdots$ respectively, the domain between two polygons is named a "layer" (Fig.2). Afterwards each layer is further divided into elements. It may be proceeded in such a way: some points on Γ_0 are selected as nodes, where the vertices of Γ_0 must be nodes and some other nodes on the line segments can be properly selected according to our requirement, then rays are drawn from the origin to the nodes, consequently every layer is divided into some quadrilaterals which are similar to each other among the layers, finally each quadrilateral is divided into two triangles such that the manner of partition for each layer is the same (Fig.2).

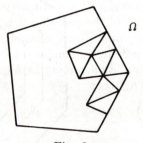

Fig. 3

Sometimes the boundary Γ_0 may not be convex, but a polygon in very complicated shape. Then we may decompose the exterior domain of Γ_0 to an unbounded domain with regular shape and a bounded domain with complicated shape. We still denote by Ω the unbounded domain, which can be divided in the above way, while the bounded domain can be divided into finite elements in conventional way (Fig.3). We need the two styles of partition to be conform to each other, that is the nodes and line segments of the elements are coincide to each other along the interior boundary. We still first analyse the domain Ω for this situation. We will give an algorithm for calculating the combined stiffness matrix of Ω, which is a matrix with finite order, and is applied either to solve a boundary value problem directly, or to solve a boundary value problem for a domain with complicated shape by assembling it with the stiffness matrices of other elements. For the later case Ω is treated as one element.

Now we consider the polygon Γ_k, and starting from one node, we arrange the nodes of Γ_k in an order according to the anticlockwise direction. Let the number of nodes be n. The nodal values of the solution of (1.1.1) are also arranged in an order, which are denoted by $y_k^{(1)}, y_k^{(2)}, \cdots, y_k^{(n)}$, and they form a n-dimensional column vector $y_k = (y_k^{(1)}, y_k^{(2)}, \cdots, y_k^{(n)})^T$, where by T we refer to the transpose of a matrix or a vecter.

1.1 2-D LAPLACE EQUATION

We consider the k-th layer between polygons Γ_{k-1} and Γ_k, the element stiffness matrix of each triangular element can be evaluated by conventional way if linear interpolation for nodal values is applied (for instance, see [61]). Upon assembling the element stiffness matrices by the nodes in the k-th layer, we obtain a stiffness matrix of one layer, which is denoted by

$$\begin{pmatrix} K_0 & -A^T \\ -A & K_0' \end{pmatrix}, \tag{1.1.5}$$

that is, if the strain energy of the k-th layer is W_k, then

$$W_k = \frac{1}{2} \begin{pmatrix} y_{k-1}^T & y_k^T \end{pmatrix} \begin{pmatrix} K_0 & -A^T \\ -A & K_0' \end{pmatrix} \begin{pmatrix} y_{k-1} \\ y_k \end{pmatrix},$$

where K_0, K_0', and A are $n \times n$ matrices. It is known by the fundamental theory of the finite elemet method that (1.1.5) is a symmetric matrix, therefore A^T and A are the transpose of each other, and K_0, K_0' are symmetric.

By (1.1.4) and the similarity of the partition it is easy to see that the stiffness matrices for all layers are the same. Upon assembling the layer stiffness matrices by the nodes, we obtain an infinite by infinite total stiffness matrix. Noting the equation and the boundary condition, we get an algebraic system of equations of infinite order. Let $K = K_0 + K_0'$, then

$$K_0 y_0 - A^T y_1 = f_0, \tag{1.1.6}_0$$

$$-A y_0 + K y_1 - A^T y_2 = 0, \tag{1.1.6}_1$$

$$\cdots\cdots,$$

$$-A y_{k-1} + K y_k - A^T y_{k+1} = 0, \tag{1.1.6}_k$$

$$\cdots\cdots.$$

For the problem (1.1.1) (1.1.2) y_0 is known and there is no equation on Γ_0, so we should drop the equation $(1.1.6)_0$. For the problem (1.1.1) (1.1.3) we need the equation $(1.1.6)_0$ on Γ_0, where f_0 is the "load vecter", which can be evaluated from the boundary value f by conventional finite element method. Now our aim is giving the algorithm of the combined stiffness matrix on Ω, thus we ignore the difference between these two boundary conditions at this occasion.

We will prove in the next chapter that there exists a real $n \times n$ matrix X such that

$$y_{k+1} = X y_k, \quad k = 0, 1, \cdots. \tag{1.1.7}$$

X is defined as a transfer matrix. Substituting (1.1.7) into $(1.1.6)_1$ we obtain

$$(-A + KX - A^T X^2) y_0 = 0.$$

y_0 is an arbitrary vector, hence X satisfies the equation

$$A^T X^2 - KX + A = 0. \tag{1.1.8}$$

Let
$$R_1 = \begin{pmatrix} K & -A \\ I & 0 \end{pmatrix}, \quad R_2 = \begin{pmatrix} A^T & 0 \\ 0 & I \end{pmatrix}, \tag{1.1.9}$$
where I is the unit matrix. The equation $(1.1.6)_k$ can be written as
$$R_1 \begin{pmatrix} y_k \\ y_{k-1} \end{pmatrix} = R_2 \begin{pmatrix} y_{k+1} \\ y_k \end{pmatrix}. \tag{1.1.10}$$

Let λ, g be a couple of eigenvalue and eigenvecter, then
$$Xg = \lambda g. \tag{1.1.11}$$

Taking $k = 1$ and setting $y_0 = g$ in (1.1.10), and noting (1.1.7) (1.1.11) we get
$$R_1 \begin{pmatrix} \lambda g \\ g \end{pmatrix} = \lambda R_2 \begin{pmatrix} \lambda g \\ g \end{pmatrix}. \tag{1.1.12}$$

Therefore λ and $\begin{pmatrix} \lambda g \\ g \end{pmatrix}$ are the generalized eigenvalue and generalized eigenvector of the matrices bundle $R_1 - \lambda R_2$ [43]. To find matrix X, we first solve the above generalized eigenvalue problem to get all $2n$ eigenvalues and eigenvectors, then pick out the n eigenvalues and eigenvectors of X.

We will always denote by det the determinant of a matrix. Now we evaluate
$$\det(R_1 - \lambda R_2) = \det \begin{pmatrix} K - \lambda A^T & -A \\ I & -\lambda I \end{pmatrix}.$$

This matrix is split into blocks. We add the second colomn by the product of the first column and λ, then obtain
$$\det(R_1 - \lambda R_2) = \det \begin{pmatrix} K - \lambda A^T & -A + \lambda K - \lambda^2 A^T \\ I & 0 \end{pmatrix}$$
$$= (-1)^n \det(-A + \lambda K - \lambda^2 A^T).$$

K is symmetric, so this expression is a symmetric polynomial with the independent variable λ. Therefore if $\lambda \neq 0$ is an eigenvalue, then so is $1/\lambda$. We will prove in the next chapter that the eigenvalues of X satisfy $|\lambda| \leq 1$, and if $|\lambda| = 1$ then necessarily $\lambda = 1$ which corresponds to the eigenvector $g_1 = (1, 1, \cdots, 1)^T$. Therefore we only need to pick out the eigenvalues, which satisfy $|\lambda| < 1$, of the matrices bundle $R_1 - \lambda R_2$, then supply a $\lambda = 1$. Let the eigenvalues be $\lambda_1, \lambda_2, \cdots, \lambda_n$, and the corresponding eigenvectors be $g_1, g_2, \cdots, g_n$, and we set $T = (g_1, g_2, \cdots, g_n)$, $\Lambda = \text{diag}(\lambda_1, \lambda_2, \cdots, \lambda_n)$, where diag means a diagonal matrix or a block diagonal matrix. By (1.1.11) we have
$$XT = T\Lambda,$$

that is
$$X = T\Lambda T^{-1}. \tag{1.1.13}$$
(1.1.13) is the formula for evaluating the transfer matrix X. Substituting (1.1.7) into $(1.1.6)_0$ we obtain
$$K_z y_0 = f_0, \tag{1.1.14}$$
where $K_z = K_0 - A^T X$, which is defined as a combined stiffness matrix. We will prove that K_z is a symmetric and semi-positive definite matrix in the next chapter.

We should note that the combined stiffness matrix is different from the total stiffness matrix. At this circumstance, the total stiffness matrix is an infinite by infinite matrix, by which we can get the strain energy arised from any nodal values, but the combined stiffness matrix is only a $n \times n$ matrix, by which we can only get the strain energy when the nodal values $y_0, y_1, \cdots$ satisfy the equations $(1.1.6)_1, (1.1.6)_2, \cdots$. The procedure from the the total stiffness matrix to the combined stiffness matrix can be viewed as a procedure of elimination. By this procedure, $y_1, y_2, \cdots$ are all eliminated.

The procedure of solving problems by the infinite element method is the following: First of all we find the transfer matrix and combined stiffness matrix, then solve an algebraic system in analogues with the finite element method. Once y_0 is obtained, the other nodal values can be determined by (1.1.7).

If A is invertible, then the above generalized eigenvalue problem can be reduced to a conventional eigenvalue problem. Multiplying R_2^{-1} on the left of (1.1.12), we get
$$\begin{pmatrix} (A^T)^{-1} K & -(A^T)^{-1} A \\ I & 0 \end{pmatrix} \begin{pmatrix} \lambda g \\ g \end{pmatrix} = \lambda \begin{pmatrix} \lambda g \\ g \end{pmatrix},$$
which is a conventional eigenvalue problem.

Eigenvalue λ may be a complex number, thus the above calculation may encounter complex numbers. To prevent from this inconvenience, we can vary (1.1.13) slightly. Eigenvalues and eigenvectors must appear in conjugate pairs. Let $\lambda = \alpha \pm i\beta, g = p \pm iq$ be a pair of eigenvalues and eigenvectors, then by (1.1.11) we have
$$X(p \pm iq) = (\alpha \pm i\beta)(p \pm iq).$$
Separating the real and imaginary parts we obtain
$$Xp = \alpha p - \beta q, \quad Xq = \beta p + \alpha q,$$
that is
$$X \begin{pmatrix} p & q \end{pmatrix} = \begin{pmatrix} p & q \end{pmatrix} \begin{pmatrix} \alpha & \beta \\ -\beta & \alpha \end{pmatrix}.$$
We substitute the corresponding two columns in matrices T and Λ, and let
$$T_1 = (\cdots, p, q, \cdots),$$
$$\Lambda_1 = \text{diag}\left(\cdots, \begin{pmatrix} \alpha & \beta \\ -\beta & \alpha \end{pmatrix}, \cdots\right).$$

It is easy to see

$$X = T_1 \Lambda_1 T_1^{-1}.$$

After substituting with respect to all complex eigenvalues, the calculation of the transfer matrix X becomes real.

1.2 Fourier method

The method given in the previous section is effective for any similar partition, and the amount of calculation is to solve a $2n \times 2n$ matrix eigenvalue problem. In this section we apply the Fourier transform for a particular similar partition, and only a very little amount of calculation is needed to find the matrices X and K_z.

We require that the partition in Fig.2 admits the following property: If we rotate this figure $\frac{2\pi}{n}$ radian around point O, then it is invariable. In that case Γ_0 is a regular polygon with centre O, and the partition of all quadrilaterals are uniform, see Fig. 4.

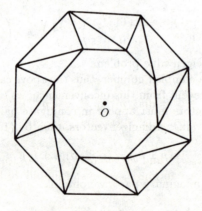

Fig. 4

We will prove in the next chapter that the matrices K_0, K_0' and A are all circulant matrices for this partition, i.e. they have the form

$$B = \begin{pmatrix} b_1 & b_2 & \cdots & b_n \\ b_n & b_1 & \cdots & b_{n-1} \\ \multicolumn{4}{c}{\dotfill} \\ b_2 & \cdots & b_n & b_1 \end{pmatrix}.$$

We denote the unit root by $\omega = e^{\frac{2\pi\sqrt{-1}}{n}}$ and we define a unitary matrix

$$F = \frac{1}{\sqrt{n}} \begin{pmatrix} 1 & 1 & \cdots & 1 \\ 1 & \omega & \cdots & \omega^{n-1} \\ 1 & \omega^2 & \cdots & \omega^{2(n-1)} \\ \cdots\cdots\cdots\cdots\cdots\cdots \\ 1 & \omega^{n-1} & \cdots & \omega^{(n-1)^2} \end{pmatrix},$$

then we can apply a Fourier transform to B as follows: Let $\bar{F}$ be the conjugate matrix of F, then it can be verified by simple calculation that

$$\bar{F}BF = \operatorname{diag}\left(\sum_{i=1}^{n} b_i, \sum_{i=1}^{n} \omega^{i-1} b_i, \cdots, \sum_{i=1}^{n} \omega^{(i-1)(n-1)} b_i\right).$$

Let $y_k = Fz_k, k = 0, 1, \cdots$, then multiplying $\bar{F}$ on the left of the equations of (1.1.6), we obtain

$$P_0 z_0 - \bar{Q} z_1 = \bar{F} f_0, \qquad (1.2.1)_0$$

$$-Q z_0 + P z_1 - \bar{Q} z_2 = 0, \qquad (1.2.1)_1$$

$$\cdots\cdots\cdots$$

$$-Q z_{k-1} + P z_k - \bar{Q} z_{k+1} = 0, \qquad (1.2.1)_k$$

$$\cdots\cdots\cdots$$

where $P_0 = \bar{F} K_0 F, Q = \bar{F} A F, P = \bar{F} K F$, all of which are diagonal matrices. We denote them by

$$P_0 = \operatorname{diag}(p_0^{(1)}, p_0^{(2)}, \cdots, p_0^{(n)}),$$
$$P = \operatorname{diag}(p^{(1)}, p^{(2)}, \cdots, p^{(n)}),$$
$$Q = \operatorname{diag}(q^{(1)}, q^{(2)}, \cdots, q^{(n)}).$$

Let $z_k = (z_k^{(1)}, z_k^{(2)}, \cdots, z_k^{(n)})^T$ and $\bar{F} f_0 = (\phi^{(1)}, \phi^{(2)}, \cdots, \phi^{(n)})^T$. Writing (1.2.1) in the form of components, we obtain n systems of infinite order ($i = 1, 2, \cdots, n$),

$$p_0^{(i)} z_0^{(i)} - \bar{q}^{(i)} z_1^{(i)} = \phi^{(i)},$$

$$-q^{(i)} z_0^{(i)} + p^{(i)} z_1^{(i)} - \bar{q}^{(i)} z_2^{(i)} = 0,$$

$$\cdots\cdots\cdots$$

$$-q^{(i)} z_{k-1}^{(i)} + p^{(i)} z_k^{(i)} - \bar{q}^{(i)} z_{k+1}^{(i)} = 0,$$

$$\cdots\cdots\cdots$$

In the following our argument is similar to that of §1.1. There exist constants $x^{(i)}$ such that

$$z_{k+1}^{(i)} = x^{(i)} z_k^{(i)}, \quad k = 0, 1, \cdots.$$

Then we get the equations for $x^{(i)}$,
$$\bar{q}^{(i)}(x^{(i)})^2 - p^{(i)}x^{(i)} + q^{(i)} = 0. \tag{1.2.2}$$

The absolute values of the roots of (1.2.2) are reciprocals to each other. We will prove in the next chapter that $|x^{(i)}| \leq 1$, and if $|x^{(i)}| = 1$, then $x^{(i)} = 1$. Therefore all we have to do is to pick out one of the two roots
$$x^{(i)} = \frac{p^{(i)} \pm \sqrt{(p^{(i)})^2 - 4|q^{(i)}|^2}}{2\bar{q}^{(i)}},$$
such that the absolute value of this one is smaller than the other. Let
$$Z = \text{diag}(x^{(1)}, x^{(2)}, \cdots, x^{(n)}),$$
then
$$z_{k+1} = Zz_k, \quad k = 0, 1, \cdots,$$
$$y_{k+1} = FZ\bar{F}y_k, \quad k = 0, 1, \cdots.$$

Comparing it with §1.1, we discover that the transfer matrix is $X = FZ\bar{F}$, then we get the combined stiffness matrix
$$K_z = K_0 - A^T FZ\bar{F}$$
immediately.

1.3 Iterative method

We return now to the arbitrary similar partition of §1.1. Although formulas (1.1.13) (1.1.14) are exact, approximate computation is still needed to solve the eigenvalue problem. Here we give another approximate algorithm, the amount of calculation of which is less than that of §1.1.

We consider a set which consists of $2^l (l = 0, 1, \cdots)$ layers, and let its combined stiffness matrix be
$$\begin{pmatrix} K_l & -A_l^T \\ -A_l & K_l' \end{pmatrix}.$$

This notation coincides to that in §1.1 as $l = 0$ provided taking $A_0 = A$. Utilizing the fact that the solution to the Laplace equation makes strain energy reach its minimum value, we can get the recurrence relation,
$$(y^T \ z^T) \begin{pmatrix} K_{l+1} & -A_{l+1}^T \\ -A_{l+1} & K_{l+1}' \end{pmatrix} \begin{pmatrix} y \\ z \end{pmatrix}$$
$$= \min_{w \in \mathbb{R}^n} \left\{ (y^T \ w^T) \begin{pmatrix} K_l & -A_l^T \\ -A_l & K_l' \end{pmatrix} \begin{pmatrix} y \\ w \end{pmatrix} + (w^T \ z^T) \begin{pmatrix} K_l & -A_l^T \\ -A_l & K_l' \end{pmatrix} \begin{pmatrix} w \\ z \end{pmatrix} \right\},$$
$$\tag{1.3.1}$$

where $\mathbb{R}^n$ is the n-dimensional real vector space. Evaluating the partial derivatives of the expression in the braces with respect to w, and setting it to be zero, we get

$$-2A_l y + 2K'_l w + 2K_l w - 2A_l^T z = 0.$$

Hence

$$w = (K_l + K'_l)^{-1}(A_l y + A_l^T z). \tag{1.3.2}$$

Substituting (1.3.2) into (1.3.1) and comparing the corresponding terms, we obtain

$$K_{l+1} = K_l - A_l^T(K_l + K'_l)^{-1}A_l, \tag{1.3.3}$$

$$K'_{l+1} = K'_l - A_l(K_l + K'_l)^{-1}A_l^T, \tag{1.3.4}$$

$$A_{l+1} = A_l(K_l + K'_l)^{-1}A_l. \tag{1.3.5}$$

By (1.3.3)–(1.3.5) we obtain the combined stiffness matrices of 2^l layers recurrently for all l.

Now, we assume that y_0 is known, c is a constant to be determined, and assume that $y_k^{(i)} = c$ for $k \geq 2^l, i = 1, 2, \cdots$, and equations $(1.1.6)_k$ hold for $k = 1, 2, \cdots, 2^l - 1$. Under these restrictions, the strain energy given by (1.1.4) is

$$W = \frac{1}{2}(y_0^T \ \ cg_1^T)\begin{pmatrix} K_l & -A_l^T \\ -A_l & K'_l \end{pmatrix}\begin{pmatrix} y_0 \\ cg_1 \end{pmatrix}$$

$$= \frac{1}{2}y_0^T K_l y_0 - cg_1^T A_l y_0 + \frac{1}{2}c^2 g_1^T K'_l g_1, \tag{1.3.6}$$

where g_1 is just the eigenvector $(1, 1, \cdots, 1)^T$ in §1.1.

We take the value of c so that (1.3.6) reaches its minimum value, that is, set $\frac{\partial w}{\partial c} = 0$, then we obtain

$$-g_1^T A_l y_0 + cg_1^T K'_l g_1 = 0.$$

Therefore

$$c = \frac{g_1^T A_l y_0}{g_1^T K'_l g_1} = \frac{y_0^T A_l^T g_1}{g_1^T K'_l g_1}.$$

Thus

$$W = \frac{1}{2}y_0^T\left(K_l - \frac{A_l^T g_1 g_1^T A_l}{g_1^T K'_l g_1}\right)y_0.$$

Let

$$K_z^{(l)} = K_l - \frac{A_l^T g_1 g_1^T A_l}{g_1^T K'_l g_1}. \tag{1.3.7}$$

We will prove in the next chapter that

$$\lim_{l \to \infty} K_z^{(l)} = K_z. \tag{1.3.8}$$

Once the combined stiffness matrix K_z is obtained, by the equation (1.1.8) we have

$$-(K_z + K_0')X + A = 0,$$

then we can find the transfer matrix

$$X = (K_z + K_0')^{-1}A.$$

We will prove in the next chapter that the rate of convergence of (1.3.8) is very high. But this iterative scheme has a considerable shortcoming. Numerical experiment has shown that probably one significant digit of the decimal scale is lost during once calculation of (1.3.3)–(1.3.5). Usually $K_z^{(l)}$ does not change along with l after a few times of iteration, but it is still not the needed combined stiffness matrix K_z. This is the so called phenomenon of "pseudo-convergence". To overcome this difficulty, we may use more digits in calculation or apply the following iterative method to correct the obtained values.

We will give another iterative scheme. The rate of convergence is not so high, but it is stable. Our suggestion is the following: First (1.3.3)–(1.3.5),(1.3.7) are applied for only a few times to get a roughly approximate matrix of K_z, and it is used as the initial data of the following iterative scheme, then a highly precise matrix K_z is obtained.

We fix $l \geq 0$ and assume that the combined stiffness matrix for 2^l layers is known and precise enough. By similarity, the combined stiffness matrix on the unbounded domain which is the exterior of this 2^l layers is still K_z, thus we have

$$y_0^T K_z y_0 = \min_{w \in \mathbb{R}^n} \left\{ \begin{pmatrix} y_0^T & w^T \end{pmatrix} \begin{pmatrix} K_l & -A_l^T \\ -A_l & K_l' \end{pmatrix} \begin{pmatrix} y_0 \\ w \end{pmatrix} + w^T K_z w \right\}. \tag{1.3.9}$$

Evaluating the partial derivatives of the expression in the braces with respect to w, and setting it to be zero, we get

$$-2A_l y_0 + 2K_l' w + 2K_z w = 0,$$

thus

$$w = (K_z + K_l')^{-1} A_l y_0.$$

Substituting it into (1.3.9), we obtain

$$y_0^T K_z y_0 = y_0^T (K_l - A_l^T (K_z + K_l')^{-1} A_l) y_0.$$

But y_0 is arbitrary, by this we obtain the matrix equation for K_z,

$$K_z = K_l - A_l^T (K_z + K_l')^{-1} A_l. \tag{1.3.10}$$

We will prove in the next chapter that g_1 is the null eigenvector (the eigenvector associated to the eigenvalue zero) of K_z. We may design the following iterative scheme according to (1.3.10):

Step 1. If a symmetric and semi-positive definite matrix $K_z^{(m)}$, which possesses the null eigenvector g_1, is known, then we set

$$K_z^{(m+1)} = K_l - A_l^T (K_z^{(m)} + K_l')^{-1} A_l. \tag{1.3.11}$$

Step 2. Correction.

Theoretically we will prove in the next chapter that $K_z^{(m+1)}$ is also a symmetric and semi-positive definite matrix, and it possesses the null eigenvector g_1, but unavoidably there is some round-off error during the calculation of (1.3.11), hence $K_z^{(m+1)}$ may not admit the above properties completely. As a result this iterative process may fail to converge. Therefore we correct $K_z^{(m+1)}$ properly. First we replace k_{ij} and k_{ji} by $\frac{1}{2}(k_{ij} + k_{ji})$ for $i \neq j$, then we set $k_{ii} = -\sum_{j \neq i} k_{ij}$.

Step 3. Replacing m by $m+1$, we return to step 1.

We will prove in the next chapter that

$$\lim_{m \to \infty} K_z^{(m)} = K_z.$$

and the "pseudo-convergence" phenomenon would not happen for this scheme provided $K_z^{(0)}$ is a symmetric and semi-positive matrix and possesses a null eigenvector g_1.

1.4 General elements

The pattern of partition in §1.1 can be generalized. On the one hand the elements are not restricted to be linear triangular elements, on the other hand each layer may not be divided by rays. We should notice that we have not made any other assumption to the stiffness matrices in the preceding argument except that they are the same for all layers. Therefore the prerceding argument is applicable for more general partitions.

An example is shown in Fig.5 where part of the partition is not constructed by rays, and both linear triangular elements and bilinear isoparametric quadrilateral elements are involved in the mesh. As a matter of fact, the partition in each layer is quite flexiable. There are only two restrictions, namely, one to one correspondence of the nodes between Γ_{k-1} and Γ_k and the geometry of the layer mesh is similar to each other. Fig. 6 shows an example of applying quadrilateral eight nodes isoparametric elements.

It should be noticed that there are some nodes which do not belong to Γ_k or Γ_{k-1} in both Fig.5 and Fig.6. It is needed to eliminate those nodal values from the systems and obtain the combined stiffness matrix for one layer. Let y be a column vector consisting of the interior nodal values, then we can write the strain energy

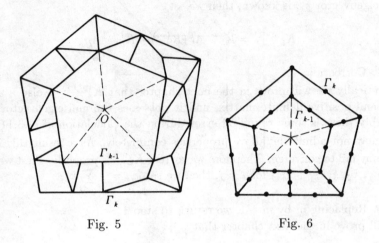

Fig. 5   Fig. 6

for one layer as

$$W = \frac{1}{2}(y_{k-1}^T \quad y^T \quad y_k^T)\begin{pmatrix} K_{11} & K_{12} & K_{13} \\ K_{21} & K_{22} & K_{23} \\ K_{31} & K_{32} & K_{33} \end{pmatrix}\begin{pmatrix} y_{k-1} \\ y \\ y_k \end{pmatrix}, \quad (1.4.1)$$

in terms of the total stiffness matrix. Letting $\frac{\partial W}{\partial y} = 0$, we get

$$K_{22}y + \frac{1}{2}(K_{12} + K_{21})y_{k-1} + \frac{1}{2}(K_{23} + K_{32})y_k = 0,$$

then we can solve

$$y = -K_{22}^{-1}\left\{\frac{1}{2}(K_{12} + K_{21})y_{k-1} + \frac{1}{2}(K_{23} + K_{32})y_k\right\}. \quad (1.4.2)$$

Upon substituting (1.4.2) into (1.4.1), and rearranging the terms, we can get

$$W = \frac{1}{2}(y_{k-1}^T \quad y_k^T)\begin{pmatrix} K_0 & -A^T \\ -A & K_0' \end{pmatrix}\begin{pmatrix} y_{k-1} \\ y_k \end{pmatrix}.$$

The combined stiffness matrices for all layers are the same.

The approach in §1.2 is applicable for more general partitions too. Fig.7 shows one example, where the mesh keeps fixed after rotating $\frac{\pi}{2}$ radians. The discussion involves block circulant matrices. We will discuss the Fourier method for block circulant matrices at the later occasion when we consider the calculation of plane stress.

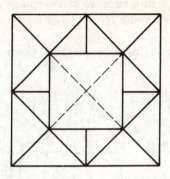

Fig. 7

1.5 Three dimensional exterior problems of the Laplace equation

The following discussion is similar to that of the two dimensional case. Let Γ_0 be a closed convex surface in a three dimensional space, it is stitched up with space quadrilaterals, and the point O is in the interior of Γ_0. Taking $\xi > 1$, we draw the similar surfaces to Γ_0 with centre O and the constants of proportionality $\xi, \xi^2, \cdots, \xi^k, \cdots$, which are denoted by $\Gamma_1, \Gamma_2, \cdots, \Gamma_k, \cdots$ respectively. The domain between Γ_{k-1} and Γ_k may be divided into many sorts of elements, for instance eight nodes hexahedron isoparametric elements. Let

$$\begin{pmatrix} K_0 & -A^T \\ -A & K_0' \end{pmatrix}$$

be the combined stiffness matrix of the layer between Γ_0 and Γ_1, then the combined stiffness matrices of the other layers are

$$\xi^{k-1} \begin{pmatrix} K_0 & -A^T \\ -A & K_0' \end{pmatrix}, \quad k = 2, 3, \cdots.$$

Let $K = \xi^{\frac{1}{2}} K_0 + \xi^{-\frac{1}{2}} K_0'$, then the equation corresponding to $(1.1.6)_k$ is

$$-A y_{k-1} + \xi^{\frac{1}{2}} K y_k - \xi A^T y_{k+1} = 0.$$

Hence the transfer matrix X satisfies the equation

$$\xi A^T X^2 - \xi^{\frac{1}{2}} K X + A = 0.$$

If we set $\xi^{\frac{1}{2}} X = Y$, then Y satisfies the equation

$$A^T Y^2 - K Y + A = 0.$$

We will prove in the next chapter that all eigenvalues λ of the transfer matrix X satisfy $|\lambda| < \xi^{-\frac{1}{2}}$, hence all eigenvalues λ of the matrix Y satisfy $|\lambda| < 1$. We define matrices R_1, R_2 by (1.1.9), then we can evaluate the general eigenvalues and eigenvectors of the matrices bundle $R_1 - \lambda R_2$. Finally we get the matrix Y by the approach in §1.1.

We turn now to consider the iterative method. Corresponding to (1.3.1) we have the equation

$$\begin{pmatrix} y^T & z^T \end{pmatrix} \begin{pmatrix} K_{l+1} & -A_{l+1}^T \\ -A_{l+1} & K'_{l+1} \end{pmatrix} \begin{pmatrix} y \\ z \end{pmatrix}$$
$$= \min_{w \in \mathbb{R}^n} \left\{ \begin{pmatrix} y^T & w^T \end{pmatrix} \begin{pmatrix} K_l & -A_l^T \\ -A_l & K'_l \end{pmatrix} \begin{pmatrix} y \\ w \end{pmatrix} \right.$$
$$\left. + \xi^{2^l} \begin{pmatrix} w^T & z^T \end{pmatrix} \begin{pmatrix} K_l & -A_l^T \\ -A_l & K'_l \end{pmatrix} \begin{pmatrix} w \\ z \end{pmatrix} \right\}.$$

Thus

$$w = (\xi^{2^l} K_l + K'_l)^{-1}(A_l y + \xi^{2^l} A_l^T z),$$
$$K_{l+1} = K_l - A_l^T(\xi^{2^l} K_l + K'_l)^{-1} A_l,$$
$$K'_{l+1} = \xi^{2^l} K'_l - \xi^{2^{l+1}} A_l(\xi^{2^l} K_l + K'_l)^{-1} A_l^T,$$
$$A_{l+1} = \xi^{2^l} A_l(\xi^{2^l} K_l + K'_l)^{-1} A_l.$$

We will prove in the next chapter that

$$\lim_{l \to \infty} K_l = K_z.$$

For the iterative method of the second type, we have

$$y_0^T K_z y_0$$
$$= \min_{w \in \mathbb{R}^n} \left\{ \begin{pmatrix} y_0^T & w^T \end{pmatrix} \begin{pmatrix} K_l & -A_l^T \\ -A^T & K'_l \end{pmatrix} \begin{pmatrix} y_0 \\ w \end{pmatrix} + \xi^{2^l} w^T K_z w \right\}$$

on the analogy of (1.3.9), thus

$$w = (\xi^{2^l} K_z + K'_l)^{-1} A_l y_0,$$
$$K_z = K_l - A_l^T(\xi^{2^l} K_z + K'_l)^{-1} A_l.$$

Finally we have an iterative scheme similar to (1.3.11),

$$K_z^{(m+1)} = K_l - A_l^T(\xi^{2^l} K_z^{(m)} + K'_l)^{-1} A_l,$$

where $K_z^{(0)}$ is a symmetric and positive definite matrix.

1.6 Problems on other unbounded domains

We consider for example the problem on a plane domain consisting of two subdomains, which are a bounded domain Ω_1 and the lower half plane $\{y < 0\}$ (Fig.8). For definiteness we first assume that the boundary condition is $\frac{\partial u}{\partial y} = 0$ on the partial boundary lying on the x-axis. Some boundary condition is also given on the other part of the boundary, then we intend to solve the boundary value problem of the Laplace equation.

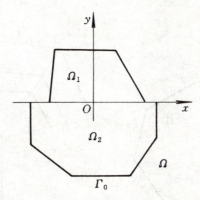

Fig. 8

We draw a broken line Γ_0 on $\{y < 0\}$ such that it begins at the negative x-axis and ends up at the positive x-axis. $\{y < 0\}$ is divided by Γ_0 into two subdomains, where the bounded one is Ω_2, and the unbounded one is Ω (Fig.8). After doing an infinite similar partition on Ω, we can apply the approaches in §1.1 and §1.3 to this case without any modification, then obtain the combined stiffness matrix K_z on the domain Ω. If the number of nodes on Γ_0 is n, then K_z is an $n \times n$ matrix.

Secondly, if the boundary condition is $u = 0$ on the part of boundary lying on the x-axis, then K_0, K_0', A are all $(n-2) \times (n-2)$ matrices after dropping the nodes on the x-axis, consequently K_z is a $(n-2) \times (n-2)$ matrix. Moreover, the eigenvalues λ of the transfer matrix X satisfy $|\lambda| < 1$. The recurrent relations (1.3.3)–(1.3.5) are also applicable for calculating K_l, K_l' and A_l. We can also prove

$$\lim_{l \to \infty} K_l = K_z,$$

the proof of which will be given in the next chapter. The iterative scheme (1.3.11) still holds, but $K_z^{(m)}$ is a symmetric and positive definite matrix now.

1.7 Corner problems

Suppose there is a corner point O on the boundary of the considered domain (bounded or unbounded) (Fig.9) with interior angle $\alpha > \pi$, then generally speaking the solution possesses singularity at the point O; or although $\alpha \leq \pi$ but different sorts of boundary conditions are given on the two adjacent sides of the point O, then generally speaking there is also a singularity at the point O. Here by "singularity" we refer to the property that the derivatives of the solutions will grow infinitely as the point tends to the point O. It is difficult to calculate this singularity by means of conventional finite element methods, moreover by those the error would propagate to the whole domain. To improve the precision of calculation, we may use the infinite element method.

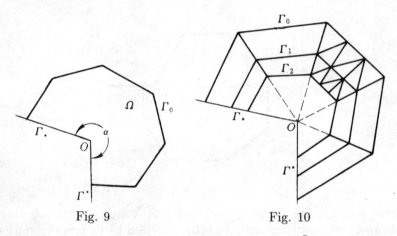

Fig. 9   Fig. 10

For definiteness, we first assume a boundary condition $\frac{\partial u}{\partial \nu} = 0$ on the adjacent sides to the point O. Then we draw a broken line around the point O which begins at one adjacent side Γ^* to the point O and ends up at another adjacent side Γ_* of it, see Fig. 9. We denote by Ω the domain surrounding by Γ^*, Γ_* and Γ_0. Taking a constant ξ such that $0 < \xi < 1$, we draw similar broken lines $\Gamma_1, \Gamma_2, \cdots, \Gamma_k, \cdots$ with centre O and the constants of proportionality $\xi, \xi^2, \cdots, \xi^k, \cdots$, then copying after the approach in §1.1, we get infinite similar mesh shown in Fig.10.

The argument follows the same lines as those of §1.1 and §1.3. In addition (1.3.7) can be replaced by a faster scheme.

Let point O be the origin, we define two new n-dimensional vectors $z_1 = (z_1^{(1)}, z_1^{(2)}, \cdots, z_1^{(n)})^T$ and $z_2 = (z_2^{(1)}, z_2^{(2)}, \cdots, z_2^{(n)})^T$, such that the components of them are the x-coordinate and the y-coordinate of the nodes accordingly. Define vectors $\xi^k z_1$ and $\xi^k z_2$, $k = 0, 1, \cdots$, on Γ_k, then the linear interpolation of them are functions x and y respectively. It is evident that they are infinite element solutions, hence z_1 and z_2 are the eigenvectors of the transfer matrix X corresponding to the eigenvalue $\lambda = \xi$.

Letting c_1, c_2, c_3 be constants to be determined, we set $u = c_1 + c_2 x + c_3 y$ on every element for $k \geq 2^l$. Let the area of domain Ω be S, then the area of the domain surrounding by Γ_{2^l}, Γ^* and Γ_* is $\xi^{2^{l+1}} S$. The strain energy on the above-mentioned domain is equal to $\frac{1}{2}\xi^{2^{l+1}}S(c_2^2+c_3^2)$, and the total strain energy on the domain Ω is

$$W = \frac{1}{2}\begin{pmatrix}y_0^T & c_1 g_1^T + c_2\xi^{2^l}z_1^T + c_3\xi^{2^l}z_2^T\end{pmatrix}\begin{pmatrix}K_l & -A_l^T \\ -A_l & K_l'\end{pmatrix}$$
$$\cdot \begin{pmatrix}y_0 \\ c_1 g_1 + c_2\xi^{2^l}z_1 + c_3\xi^{2^l}z_2\end{pmatrix} + \frac{1}{2}\xi^{2^{l+1}}S(c_2^2+c_3^2).$$

Let

$$c = \begin{pmatrix}c_1 & c_2 & c_3\end{pmatrix}^T,$$

$$E_l = \begin{pmatrix}g_1^T \\ \xi^{2^l}z_1^T \\ \xi^{2^l}z_2^T\end{pmatrix}A_l,$$

$$F_l = \begin{pmatrix}g_1^T \\ \xi^{2^l}z_1^T \\ \xi^{2^l}z_2^T\end{pmatrix}K_l'\begin{pmatrix}g_1 & \xi^{2^l}z_1 & \xi^{2^l}z_2\end{pmatrix} + \xi^{2^{l+1}}S\begin{pmatrix}0 & 0 & 0 \\ 0 & 1 & 0 \\ 0 & 0 & 1\end{pmatrix},$$

then

$$W = \frac{1}{2}y_0^T K_l y_0 - c^T E_l y_0 + \frac{1}{2}c^T F_l c, \qquad (1.7.1)$$

where E_l is a $3 \times n$ matrix, and F_l is a 3×3 symmetric matrix. Letting $\frac{\partial W}{\partial c} = 0$, we get

$$-E_l y_0 + F_l c = 0,$$

hence

$$c = F_l^{-1} E_l y_0.$$

Substituting it into (1.7.1), we obtain

$$W = \frac{1}{2}y_0^T K_l y_0 - \frac{1}{2}y_0^T E_l^T F_l^{-1} E_l y_0.$$

We set

$$K_z^{(l)} = K_l - E_l^T F_l^{-1} E_l. \qquad (1.7.2)$$

We will prove in the next chapter that

$$\lim_{l \to \infty} K_z^{(l)} = K_z.$$

If the boundary condition is $u = 0$ on the two adjacent sides to the point O, the approach is similar. Being analogous to the previous section, the eigenvalues of the transfer matrix X satisfy $|\lambda| < 1$, and

$$\lim_{l \to \infty} K_l = K_z. \qquad (1.7.3)$$

If different boundary conditions are given on the two adjacent sides to the point O, for instance $\frac{\partial u}{\partial \nu} = 0$ is given on Γ^* and $u = 0$ on Γ_*, then the approach is still the same, we also have $|\lambda| < 1$ and (1.7.3).

1.8 Plane elasticity problems

This section is devoted to the problem of plane strain, where the system of governing equations is

$$\frac{\partial \sigma_x}{\partial x} + \frac{\partial \tau_{xy}}{\partial y} = 0,$$

$$\frac{\partial \tau_{xy}}{\partial x} + \frac{\partial \sigma_y}{\partial y} = 0,$$

$$\begin{pmatrix} \sigma_x \\ \sigma_y \\ \tau_{xy} \end{pmatrix} = \begin{pmatrix} \lambda + 2\mu & \lambda & 0 \\ \lambda & \lambda + 2\mu & 0 \\ 0 & 0 & \mu \end{pmatrix} \begin{pmatrix} \frac{\partial u}{\partial x} \\ \frac{\partial v}{\partial y} \\ \frac{\partial u}{\partial y} + \frac{\partial v}{\partial x} \end{pmatrix},$$

where $\sigma_x, \sigma_y, \tau_{xy}$ denote the stresses, u, v denote the displacements, and λ, μ are the Lamé constants. The equations for the problem of plane stress are almost the same except that the constants λ, μ are different. If displacement is given on the boundary, then the boundary condition is

$$u = f_x, \quad v = f_y, \tag{1.8.1}$$

and if distributed surface load is given on the boundary, then the boundary condition is

$$\begin{aligned} \sigma_x \cos(\nu, x) + \tau_{xy} \cos(\nu, y) &= f_x, \\ \tau_{xy} \cos(\nu, x) + \sigma_y \cos(\nu, y) &= f_y. \end{aligned} \tag{1.8.2}$$

Sometimes we may have a mixed boundary condition coupling one displacement boundary condition and one surface load boundary condition on the same boundary. The strain energy on the domain is

$$W = \frac{1}{2} \int_\Omega \left\{ (\lambda + 2\mu) \left(\left(\frac{\partial u}{\partial x} \right)^2 + \left(\frac{\partial v}{\partial y} \right)^2 \right) + 2\lambda \frac{\partial u}{\partial x} \frac{\partial v}{\partial y} \right. \\ \left. + \mu \left(\frac{\partial u}{\partial y} + \frac{\partial v}{\partial x} \right)^2 \right\} dx dy. \tag{1.8.3}$$

Sometimes the problem possesses certain property of symmetry. Fig.11 is an example, where the body is symmetric with respect to the x-axis and the surface

1.8 ELASTICITY

load is the mirror reflection of itself with the x-axis. The displacement is also the mirror reflection of itself with the x-axis up to an arbitrary rigid body motion which may be added to it. To save the computer time, we can only evaluate the solution on the subdomain which lies on the upper half plane $\{y > 0\}$. By the property of symmetry there is a sliding boundary condition on the x-axis as follows:

$$v = 0, \quad \tau_{xy} = 0,$$

which belongs to the mixed boundary conditions as mentioned above.

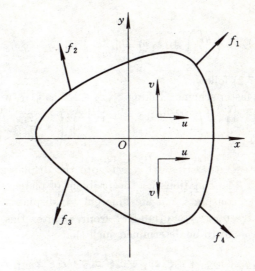

Fig. 11

The infinite element method to solve the above problem is similar to that in §1.1–§1.7. First of all we consider the corner problem (Fig.9). Let the boundary condition on the adjacent sides of the point O be load free, that is the condition (1.8.2) with $f_x = f_y = 0$.

We partition the domain in accordance with Fig.10. Let the number of nodes on Γ_k be n. According to the anticlockwise direction the displacements on the nodes are $u_k^{(1)}, v_k^{(1)}, u_k^{(2)}, v_k^{(2)}, \cdots, u_k^{(n)}, v_k^{(n)}$ successively, and they form a $2n$-dimensional column vector y_k. Now we can evaluate the stiffness matrix for one layer by (1.8.3), then the strain energy is expressed as

$$W_k = \frac{1}{2}(y_{k-1}^T \quad y_k^T)\begin{pmatrix} K_0 & -A^T \\ -A & K_0' \end{pmatrix}\begin{pmatrix} y_{k-1} \\ y_k \end{pmatrix},$$

where K_0, K_0' and A are $2n \times 2n$ matrices, and they are independent of the level k.

Let $K = K_0 + K_0'$, then the formulas (1.1.6)–(1.1.14) are all applicable, except that the property of the transfer matrix X is a little different, namely, two eigenvalues of X are equal to 1, which correspond to the eigenvectors $g_1 = (1, 0, 1, 0, \cdots, 1, 0)^T$

and $g_2 = (0,1,0,1,\cdots,0,1)^T$ respectively. They are nothing but the translational motions along the x-direction and the y-direction. All other eigenvalues satisfy $|\lambda| < 1$.

With regard to the iterative method, (1.3.3)–(1.3.5) are still applicable. To evaluate the approximate combined stiffness matrices $K_z^{(l)}$, let $y_k = c_1 g_1 + c_2 g_2$ for $k \geq 2^l$, where c_1 and c_2 are two constants to be determined. The strain energy is

$$W = \frac{1}{2} \begin{pmatrix} y_0^T & c_1 g_1^T + c_2 g_2^T \end{pmatrix} \begin{pmatrix} K_l & -A_l^T \\ -A_l & K_l' \end{pmatrix} \begin{pmatrix} y_0 \\ c_1 g_1 + c_2 g_2 \end{pmatrix}.$$

let $c = (c_1, c_2)^T$, and

$$E_l = \begin{pmatrix} g_1^T \\ g_2^T \end{pmatrix} A_l, \quad F_l = \begin{pmatrix} g_1^T \\ g_2^T \end{pmatrix} K_l' \begin{pmatrix} g_1 & g_2 \end{pmatrix}, \quad (1.8.4)$$

then (1.7.1) and (1.7.2) follow.

We may design a faster scheme. Taking $z_1, z_2 \in \mathbb{R}^n$ as §1.7, we define two vectors $z_{11} = (z_{11}^{(1)}, z_{11}^{(2)}, \cdots, z_{11}^{(2n)})^T \in \mathbb{R}^{2n}, z_{12} = (z_{12}^{(1)}, z_{12}^{(2)}, \cdots, z_{12}^{(2n)})^T \in \mathbb{R}^{2n}$ as follows $(i = 1, 2, \cdots, n)$:

$$z_{11}^{(2i-1)} = z_1^{(i)}, \quad z_{12}^{(2i-1)} = 0,$$
$$z_{11}^{(2i)} = 0, \quad z_{12}^{(2i)} = z_1^{(i)}.$$

And we define two vectors $z_{21}, z_{22} \in \mathbb{R}^{2n}$ from z_2 after this manner. We take constants c_i $(i = 1, \cdots, 6)$ to be determined such that

$$y_k = c_1 g_1 + c_2 g_2 + c_3 \xi^k z_{11} + c_4 \xi^k z_{12} + c_5 \xi^k z_{21} + c_6 \xi^k z_{22}$$

for $k \geq 2^l$. As a matter of fact we have assumed that the displacements u and v are linear functions for $k \geq 2^l$. Moreover we assume that the equations $(1.1.6)_k$ hold for $k = 1, 2, \cdots, 2^{l-1}$. By (1.8.3) the strain energy on Ω is

$$W = \frac{1}{2} \begin{pmatrix} y_0^T & y_{2^l}^T \end{pmatrix} \begin{pmatrix} K_l & -A_l^T \\ -A_l & K_l' \end{pmatrix} \begin{pmatrix} y_0 \\ y_{2^l} \end{pmatrix}$$
$$+ \xi^{2^{l+1}} S\{(\lambda + 2\mu)(c_3^2 + c_6^2) + 2\lambda c_3 c_6 + \mu(c_4 + c_5)^2\},$$

where S is the area of the domain Ω as §1.7. Set $c = (c_1, \cdots, c_6)^T$, and

$$E_l = G^T A_l, \quad (1.8.5)$$

$$F_l = G^T K_l' G + \xi^{2^{l+1}} S \begin{pmatrix} 0 & 0 & 0 & 0 & 0 & 0 \\ 0 & 0 & 0 & 0 & 0 & 0 \\ 0 & 0 & \lambda + 2\mu & 0 & 0 & \lambda \\ 0 & 0 & 0 & \mu & \mu & 0 \\ 0 & 0 & 0 & \mu & \mu & 0 \\ 0 & 0 & \lambda & 0 & 0 & \lambda + 2\mu \end{pmatrix},$$

where $G = (g_1, g_2, \xi^{2^l} z_{11}, \xi^{2^l} z_{12}, \xi^{2^l} z_{21}, \xi^{2^l} z_{22})$, then (1.7.1) and (1.7.2) follow.

Formula (1.3.11) is also applicable for this case. But there is somthing different in the step of correction. Now K_z is a symmetric and semi-positive definite matrix and g_1, g_2 are the null eigenvectors of it. We may make the following correction: Let k_{ij} be the elements of the matrix $K_z^{(m+1)}$, then we set $k_{i,2n} = -\sum_{l=1}^{n-1} k_{i,2l}$, $k_{i,2n-1} = -\sum_{l=1}^{n-1} k_{i,2l-1}$ for $1 \leq i \leq 2n$, then k_{ij} and k_{ji} are replaced by $\frac{1}{2}(k_{ij} + k_{ji})$ $(1 \leq i, j \leq 2n)$.

The above argument can be generalized to the exterior problems and free load problems on the unbounded domains of §1.6. The formulas (1.1.6)–(1.1.14) are still applicable now. And the formulas (1.3.3)–(1.3.5) and (1.8.4) (1.7.2) are also applicable for the iterative method. But the formula (1.8.5) is unapplicable, so is (1.3.11).

If the boundary condition on the adjacent sides of the point O is the fixed boundary condition $u = v = 0$ for the corner problem, we indicate the difference between it with the load free condition. Since four degrees of freedom along the boundary are dropped, now K_0, K_0' and A are all $(2n-4) \times (2n-4)$ matrices, and all eigenvalues λ of the transfer matrix X satisfy $|\lambda| < 1$. (1.3.3)–(1.3.5) are still applicable for the iterative method. Moreover we can prove

$$\lim_{l \to \infty} K_l = K_z,$$

while the calculation of $K_z^{(l)}$ is impossible and unnecessary. The formula (1.3.11) is also applicable for this case. But now K_z is a symmetric and positive definite matrix, so we only need to replace k_{ij} and k_{ji} by $\frac{1}{2}(k_{ij} + k_{ji})$ in the correction step.

The boundary conditions are various, and on the two adjacent sides the boundary conditions may be given differently. Therefore there is a great variety of the combinations of the boundary conditions. The properties of the matrices X, K_z and the iterative schemes vary with the boundary conditions. To treat these problems generally, we introduce the concept of "admissible translational motion".

The possible translational motion of an elastic body under the restriction of local boundary condition is named an admissible translational motion. For example, if load free boundary condition is given on the two adjacent sides to a corner point, then the admissible translational motion is $u \equiv c_1, v \equiv c_2$, where c_1 and c_2 are arbitrary constants; if fixed boundary condition is given on the two sides then the admissible translational motion is $u \equiv v \equiv 0$; see Fig.12, if load free boundary condition is given on the segment OA, and a sliding boundary condition,

$$-u\sin\theta + v\cos\theta = 0,$$
$$(\sigma_y - \sigma_x)\sin\theta\cos\theta + \tau_{xy}(\cos^2\theta - \sin^2\theta) = 0,$$

is given on the segment OB which possesses an inclination of θ, then the admissible translational motion is

$$u = c\cos\theta, \quad v = c\sin\theta,$$

where c is an arbitrary constant.

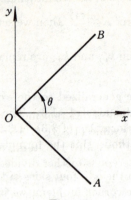

Fig. 12

We will prove in the next chapter that if the admissible translational motion possesses two degrees of freedom, then there are two eigenvalues $\lambda = 1$ of the transfer matrix X, the corresponding eigenvectors are g_1 and g_2, and all other eigenvalues satisfy $|\lambda| < 1$; if the admissible translational motion possesses one degree of freedom, then there is only one eigenvalue $\lambda = 1$, the corresponding eigenvector varies with the admissible translational motion, for the example given in Fig.12, the eigenvector is $g_1\cos\theta + g_2\sin\theta$. Those eigenvectors of the matrix X are the null eigenvectors of the combined stiffness matrix K_z at the same time.

We will also prove in the next chapter that if all eigenvalues of the transfer matrix X satisfy $|\lambda| < 1$, then we have

$$\lim_{l \to \infty} K_l = K_z \qquad (1.8.6)$$

for the iterative scheme (1.3.3)–(1.3.5). But if there is an eigenvalue $\lambda = 1$ of the matrix X, then the limit relation (1.8.6) does not hold, consequently $K_z^{(l)}$ should be evaluated by means of the approach in §1.3, where y_k are equal to the eigenvector corresponding to $\lambda = 1$ of the matrix X for $k \geq 2^l$. So as far as the corner problem is concerned, to improve the rate of convergence, we may introduce some linear functions as the additional degrees of freedom in the domain, the restriction of which is that these linear functions should satisfy the displacement boundary condition on the adjacent side (if there is any displacement boundary condition).

Let us sum up the infinite element method for the plane elasticity problems by tables. The relationship between the degrees of freedom and the properties of the transfer matrix X and the combined stiffness matrix K_z is shown in Table 1.

Table 1

degrees of freedom	multiplicity of the eigenvalue $\lambda = 1$ of X	K_z
2	2	symmetric semi-positive definite
1	1	symmetric semi-positive definite
0	0	symmetric positive definite

Table 2 is about the applicability of different methods. The method stated in §1.1 is called the eigenvalue method, while (1.3.3)–(1.3.5) in §1.3 is called the iterative method of the first type, (1.3.11) is called the iterative method of the second type. The iterative method of the first type is further divided into three subtypes, namely, limit relation (1.9.6) holds for the first subtype; (1.3.8) holds for the second subtype if some admissible translational motion is introduced; and there are faster schemes by introducing some linear functions for the third subtype. Three kinds of boundary value problems are considered, namely, the first one includes corner problems, the second one includes the problems on unbounded domains considered in §1.6, the third one includes exterior problems considered in §1.1. The sign "+" indicates the applicability of a certain method, while the sign "−" indicates unapplicable.

Table 2

	degrees of freedom	eigenvalue method	iterative method (first type)			iterative method (second type)
			I	II	III	
corner problems	2	+	−	+	+	+
	1	+	−	+	+	+
	0	+	+	−	−	+
unbounded domains	2	+	−	+	−	−
	1	+	−	+	−	−
	0	+	+	−	−	+(*)
exterior problems	2	+	−	+	−	−

(*)only for the boundary condition $u = 0$.

We conclude this section with the Fourier method of the exterior problems. We make a partition as Fig.4. The polar coordinates are applied, and at each point let u_r and u_θ be the components of displacement of the radius and anticlockwise circular directions respectively, and f_r, f_θ be the load components along these directions (Fig.13). We can get the stiffness matrix for one layer with respsct to these variables. If there are n nodes on one circle, and they are equally distributed at angles $\theta = 0, \frac{2\pi}{n}, \cdots, \frac{2(n-1)\pi}{n}$, then

$$W_k = \frac{1}{2} \begin{pmatrix} y_{k-1}^T & y_k^T \end{pmatrix} \begin{pmatrix} K_0 & -A^T \\ -A & K_0' \end{pmatrix} \begin{pmatrix} y_{k-1} \\ y_k \end{pmatrix},$$

where K_0, K_0' and A are $2n \times 2n$ matrices, and y_k are $2n$-dimensional column vectors with components u_r, u_θ on Γ_k.

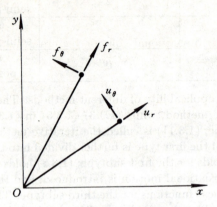

Fig. 13

We will prove in the next chapter that K_0, K_0' and A are all block circulant matrices, i.e. they are in the form of

$$B = \begin{pmatrix} b_1 & b_2 & \cdots & b_n \\ b_n & b_1 & \cdots & b_{n-1} \\ \multicolumn{4}{c}{\dotfill} \\ b_2 & \cdots & b_n & b_1 \end{pmatrix},$$

where $b_i (i = 1, \cdots, n)$ are 2×2 submatrices. We define a unitary matrix

$$F = \frac{1}{\sqrt{n}} \begin{pmatrix} I & I & \cdots & I \\ I & \omega I & \cdots & \omega^{n-1} I \\ I & \omega^2 I & \cdots & \omega^{2(n-1)} I \\ \multicolumn{4}{c}{\dotfill} \\ I & \omega^{n-1} I & \cdots & \omega^{(n-1)^2} I \end{pmatrix},$$

where I is a 2×2 unit matrix, then we can get a block diagonal matrix

$$\bar{F} B F = \text{diag} \left(\sum_{i=1}^n b_i, \sum_{i=1}^n \omega^{i-1} b_i, \cdots, \sum_{i=1}^n \omega^{(i-1)(n-1)} b_i \right)$$

by analogy with §1.2. Letting $y_k = F z_k$, and multiplying $\bar{F}$ on the left of the equations of (1.1.6), we obtain (1.2.1). The difference is that P_0, P and Q are all block diagonal matrices,

$$P_0 = \text{diag}(P_0^{(1)}, \cdots, P_0^{(n)}),$$
$$P = \text{diag}(P^{(1)}, \cdots, P^{(n)}),$$
$$Q = \text{diag}(Q^{(1)}, \cdots, Q^{(n)})$$

1.8 ELASTICITY

now, where $P_0^{(i)}, P^{(i)}$ and $Q^{(i)}$ are all 2×2 submatrices. Set

$$z_k = \begin{pmatrix} z_k^{(1)} \\ \vdots \\ z_k^{(n)} \end{pmatrix}, \quad \bar{F}f_0 = \begin{pmatrix} \varphi^{(1)} \\ \vdots \\ \varphi^{(n)} \end{pmatrix},$$

where $z_k^{(i)}, \varphi^{(i)}$ are two dimensional column vectors. (1.2.1) is rearranged into n systems with infinite order $(i = 1, \cdots, n)$,

$$\begin{aligned} P_0^{(i)} z_0^{(i)} - \bar{Q}^{(i)T} z_1^{(i)} &= \varphi^{(i)}, \\ -Q^{(i)} z_0^{(i)} + P^{(i)} z_1^{(i)} - \bar{Q}^{(i)T} z_2^{(i)} &= 0, \\ \cdots\cdots\cdots\cdots \\ -Q^{(i)} z_{k-1}^{(i)} + P^{(i)} z_k^{(i)} - \bar{Q}^{(i)T} z_{k+1}^{(i)} &= 0, \\ \cdots\cdots\cdots\cdots \end{aligned} \tag{1.8.7}$$

There exist 2×2 matrices $X^{(i)} (i = 1, 2, \cdots, n)$ such that

$$z_{k+1}^{(i)} = X^{(i)} z_k^{(i)}, \quad k = 0, 1, \cdots.$$

$X^{(i)}$ satisfy the equations

$$\bar{Q}^{(i)T}(X^{(i)})^2 - P^{(i)} X^{(i)} + Q^{(i)} = 0.$$

We will prove it in the next chapter that $X^{(i)} = I$ and all the eigenvalues λ of $X^{(i)} (2 \leq i \leq n)$ satisfy $|\lambda| < 1$. We consider the generalized eigenvalue problems of matrices bundles,

$$\begin{pmatrix} P^{(i)} & -Q^{(i)} \\ I & 0 \end{pmatrix} - \lambda \begin{pmatrix} \bar{Q}^{(i)T} & 0 \\ 0 & I \end{pmatrix}.$$

Applying the approach in §1.1 we can find $X^{(i)}$. Since the above matrices are only 4×4 ones, the amount of computer work is quite small. But the computation of complex numbers is needed now.

Let $Z = \text{diag}(X^{(1)}, X^{(2)}, \cdots, X^{(n)})$, then

$$\begin{aligned} z_{k+1} &= Z z_k, \quad k = 0, 1, \cdots, \\ y_{k+1} &= F Z \bar{F} y_k, \quad k = 0, 1, \cdots. \end{aligned}$$

By this we obtain the combined stiffness matrix $K_z = K_0 - A^T F Z \bar{F}$.

Making it easier to link the infinite elements up other elements, we transfer the coordinates to Cartesian ones for y_0 and f_0. Let the polar coordinates of a node be (r, θ), then

$$\begin{pmatrix} u \\ v \end{pmatrix} = \begin{pmatrix} \cos\theta & -\sin\theta \\ \sin\theta & \cos\theta \end{pmatrix} \begin{pmatrix} u_r \\ u_\theta \end{pmatrix} \equiv T(\theta) \begin{pmatrix} u_r \\ u_\theta \end{pmatrix}.$$

There is a same relation for the load vector. Therefore if we denote the displacement vector and the load vector in the Cartesian coordinates by y_0^* and f_0^*, and set

$$T^* = \mathrm{diag}\left(T(0), T\left(\frac{2\pi}{n}\right), \cdots, T\left(\frac{2(n-1)\pi}{n}\right)\right),$$

then
$$y_0^* = T^* y_0, \quad f_0^* = T^* f_0.$$

From $K_z y_0 = f_0$ we get
$$K_z^* y_0^* = f_0^*,$$

where
$$K_z^* = T^* K_z (T^*)^{-1}.$$

1.9 Calculation of stress intensity factors

If the interior angle is $\alpha = 2\pi$ and a local load free boundary condition is assumed for the plane elasticity corner problem, then a crack is formed (Fig.14). Let the point O be the crack tip. The stress will tend to the infinity as a moving point tends to the point O.

For definiteness, we put the crack on the positive x-axis. Applying polar coordinates, the stress intensity facter of the first type is defined as

$$K_\mathrm{I} = \lim_{r \to 0} \sqrt{2\pi r} \sigma_y(r, \pi), \tag{1.9.1}$$

and the stress intensity factor of the second type is

$$K_\mathrm{II} = \lim_{r \to 0} \sqrt{2\pi r} \tau_{xy}(r, \pi). \tag{1.9.2}$$

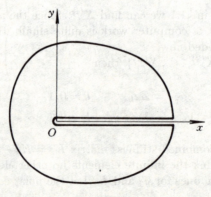

Fig. 14

1.9 STRESS INTENSITY FACTORS

Owing to the theory of elasticity mechanics [51], the stresses and displacements in the neighborhood of the point O can be developed as

$$\begin{pmatrix} \sigma_x \\ \sigma_y \\ \tau_{xy} \end{pmatrix} = \sum_{j=1}^{\infty} r^{\frac{j}{2}-1}[a_1^{(j)}\varphi_1^{(j)}(\theta) + a_2^{(j)}\varphi_2^{(j)}(\theta)], \qquad (1.9.3)$$

$$\begin{pmatrix} u \\ v \end{pmatrix} = \sum_{j=1}^{\infty} r^{\frac{j}{2}}[a_1^{(j)}\psi_1^{(j)}(\theta) + a_2^{(j)}\psi_2^{(j)}(\theta)], \qquad (1.9.4)$$

where the term of $j = 0$ corresponds to translational motion, and both $\psi_1^{(0)}(\theta)$ and $\psi_2^{(0)}(\theta)$ are constants independent of θ. The most important term in the above series is the one with index $j = 1$, where $\psi_1^{(1)}(\theta)$ corresponds to the symmetric component of displacement and $\psi_2^{(1)}(\theta)$ corresponds to the antisymmetric one, which are equal to ($0 \le \theta \le 2\pi$)

$$\psi_1^{(1)}(\theta) = \begin{pmatrix} -\frac{1}{\sqrt{8\pi}\mu}\sin\frac{\theta}{2}(\kappa - 1 + 2\cos^2\frac{\theta}{2}) \\ \frac{1}{\sqrt{8\pi}\mu}\cos\frac{\theta}{2}(\kappa + 1 - 2\sin^2\frac{\theta}{2}) \end{pmatrix}, \qquad (1.9.5)$$

$$\psi_2^{(1)}(\theta) = \begin{pmatrix} \frac{1}{\sqrt{8\pi}\mu}\cos\frac{\theta}{2}(\kappa + 1 + 2\sin^2\frac{\theta}{2}) \\ \frac{1}{\sqrt{8\pi}\mu}\sin\frac{\theta}{2}(\kappa - 1 - 2\cos^2\frac{\theta}{2}) \end{pmatrix}, \qquad (1.9.6)$$

respectively, where κ is a constant, and $\kappa = 3 - 4\nu$ for the plane strain and $\kappa = (3-\nu)(1+\nu)$ for the plane stress, where ν is the Poisson ratio. The distribution of stress for $j = 1$ is

$$\varphi_1^{(1)}(\theta) = \begin{pmatrix} \frac{1}{\sqrt{2\pi}}\sin\frac{\theta}{2}(1 + \cos\frac{\theta}{2}\cos\frac{3\theta}{2}) \\ \frac{1}{\sqrt{2\pi}}\sin\frac{\theta}{2}(1 - \cos\frac{\theta}{2}\sin\frac{3\theta}{2}) \\ \frac{1}{\sqrt{2\pi}}\sin\frac{\theta}{2}\cos\frac{\theta}{2}\sin\frac{3\theta}{2} \end{pmatrix},$$

$$\varphi_2^{(1)}(\theta) = \begin{pmatrix} \frac{1}{\sqrt{2\pi}}\cos\frac{\theta}{2}(2 - \sin\frac{\theta}{2}\sin\frac{3\theta}{2}) \\ \frac{1}{\sqrt{2\pi}}\sin\frac{\theta}{2}\cos\frac{\theta}{2}\sin\frac{3\theta}{2} \\ \frac{1}{\sqrt{2\pi}}\sin\frac{\theta}{2}(1 + \cos\frac{\theta}{2}\cos\frac{3\theta}{2}) \end{pmatrix}.$$

Upon comparing it with (1.9.1),(1.9.2) we can find that $a_1^{(1)} = K_{\mathrm{I}}$, and $a_2^{(1)} = K_{\mathrm{II}}$ exactly.

To make the calculation of stress intensity factors easier, we apply a mesh which is symmetric with respect to the x-axis as Fig.15. y_0 can be evaluated by using the approach in §1.8. Let the eigenvalues of the transfer matrix X be $\lambda_1, \lambda_2, \cdots$ arranged in an order from the greater absolute value to the smaller one. Here $\lambda_1 = \lambda_2 = 1$, the corresponding eigenvectors of which are $g_1 = (1, 0, 1, 0, \cdots, 1, 0)^T$ and $g_2 = (0, 1, 0, 1, \cdots, 0, 1)^T$ associated with translational motion. According to

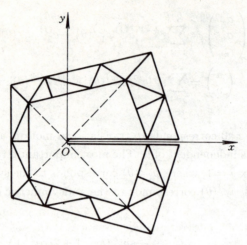

Fig. 15

the theory to be proved in Chapter 3, $\lambda_3 \approx \xi^{\frac{1}{2}}, \lambda_4 \approx \xi^{\frac{1}{2}}, \lambda_5 = \lambda_6 = \xi$. Now we explain the mechanical background of these terms.

Let the corresponding eigenvector of λ_3 be g_3, and we assume that it is the displacement vector on Γ_0, then the displacement vectors on similar polygons are $g_3, \lambda_3 g_3, \lambda_3^2 g_3, \cdots, \lambda_3^k g_3, \cdots$. We draw a ray from the point O to a node on Γ_0. This ray intersects every polygon Γ_k at a certain node because of similarity. The distance from the point O to the series of nodes is in direct proportion to ξ^k, thus

$$\text{displacement} \sim \lambda_3^k = (\xi^k)^\alpha \sim r^\alpha,$$

where $\alpha = \log\lambda_3/\log\xi \approx \frac{1}{2}$. Therefore λ_3, λ_4 are associated with the very term $j = 1$ in (1.9.4). By the same reason λ_5, λ_6 are associated with $j = 2$, which are a rotation around the point O and a simple stretch along the x-axis.

According to the theory in Chapter 3, there are an eigenvector g_3 corresponding to $\psi_1^{(1)}(\theta)$, and another eigenvector g_4 corresponding to $\psi_2^{(1)}(\theta)$. In fact, we can make g_3 symmetric and g_4 antisymmetric with respect to the displacement by taking linear combination, that is $(i = 0, 1, \cdots)$

$$g_3^{(2i+1)} = g_3^{(2n-2i-1)}, \quad g_3^{(2i+2)} = -g_3^{(2n-2i)},$$
$$g_4^{(2i+1)} = -g_4^{(2n-2i-1)}, \quad g_4^{(2i+2)} = g_4^{(2n-2i)}.$$

If g_3 and g_4 do not satisfy the above condition, then there must be one neither symmetric nor antisymmetric, and let it be g_3. By the symmetry of our mesh,

$$g_3^* = (g_3^{(2n-1)}, -g_3^{(2n)}, g_3^{(2n-3)}, -g_3^{(2n-2)}, \cdots, g_3^{(1)}, -g_3^{(2)})$$

is also an eigenvector. Thus $g_3 + g_3^*$ and $g_3 - g_3^*$ are the desired eigenvectors.

We decompose y_0 in accordance with the eigenvectors,

$$y_0 = a_1 g_1 + a_2 g_2 + \cdots + a_{2n} g_{2n}, \tag{1.9.7}$$

then the strain energy corresponding to K_I is

$$W_\text{I} = \frac{1}{2}(a_3 g_3)^T K_z (a_3 g_3), \tag{1.9.8}$$

and the one corresponding to K_II is

$$W_\text{II} = \frac{1}{2}(a_4 g_4)^T K_z (a_4 g_4). \tag{1.9.9}$$

On the other hand, let the Cartesian coordinates of the nodes on Γ_0 be (x_i, y_i), the polar coordinates of them be $(r_i, \theta_i)(i = 1, 2, \cdots, n)$, and the polar coordinates of vectors $(x_{i+1} - x_i, y_{i+1} - y_i)$ be (s_i, α_i), then we can also evaluate W_I and W_II from (1.8.3),(1.9.4)–(1.9.6) by integrating. For instance, at the case of plane strain

$$W_\text{I} = \frac{K_\text{I}^2}{8\pi\mu} \sum_{i=1}^{n-1} \frac{\begin{vmatrix} x_i & y_i \\ x_{i+1} & y_{i+1} \end{vmatrix}}{s_i} \left\{ (1 - 2\nu + \frac{1}{2}\sin^2\alpha_i)\log\frac{1 - \cos(\alpha_i - \theta_i)}{1 - \cos(\alpha_i - \theta_{i+1})} \right.$$

$$+ \frac{1}{2}(3 - 4\nu - \cos\alpha_i)(1 + \cos\alpha_i)\log\frac{r_i}{r_{i+1}} - (1 - 2\nu)(\theta_{i+1} - \theta_i)\sin\alpha_i$$

$$\left. + \frac{1}{2}\cos(\alpha_i + \theta_{i+1}) - \frac{1}{2}\cos(\alpha_i + \theta_i) \right\}, \tag{1.9.10}$$

$$W_\text{II} = \frac{K_\text{II}^2}{8\pi\mu} \sum_{i=1}^{n-1} \frac{\begin{vmatrix} x_i & y_i \\ x_{i+1} & y_{i+1} \end{vmatrix}}{s_i} \left\{ (3 - 2\nu - \frac{3}{2}\sin^2\alpha_i)\log\frac{1 - \cos(\alpha_i - \theta_i)}{1 - \cos(\alpha_i - \theta_{i+1})} \right.$$

$$+ \left[3 - 2\nu - \frac{3}{2}\sin^2\alpha_i - (1 - 2\nu)\cos\alpha_i \right] \log\frac{r_i}{r_{i+1}}$$

$$\left. + (1 - 2\nu)(\theta_{i+1} - \theta_i)\sin\alpha_i - \frac{3}{2}\cos(\alpha_i + \theta_{i+1}) + \frac{3}{2}\cos(\alpha_i + \theta_i) \right\}. \tag{1.9.11}$$

Upon comparing (1.9.8),(1.9.9) with (1.9.10),(1.9.11) we get K_I^2 and K_II^2. It should be noticed that K_I is always positive, because a negative value of K_I means that the material of two sides of the crack would intrude into each other, which has no physical meaning. However, the sign of K_II is easy to judge, since it will suffice to check the direction of the tangential displacement of $a_4 g_4$ along the crack simply.

If an elastic body is symmetric as Fig.11, then $K_\text{II} = 0$, only half of it is needed in calculation. The transfer matrix X only admits one eigenvalue $\lambda = 1$, the corresponding eigenvector is $(1, 0, 1, 0, \cdots, 1)^T$, there is only one eigenvalue $\lambda \approx \xi^{\frac{1}{2}}$ too.

We conclude this section with the way to evaluate the terms $a_3 g_3$ and $a_4 g_4$ of the expression (1.9.7). We have already known that g_3 and g_4 are the eigenvectors of the transfer matrix X and the corresponding eigenvalues λ_3 and λ_4 are approximate to $\xi^{\frac{1}{2}}$. Let $a = (a_1, a_2, \cdots, a_{2n})^T, T = (g_1, g_2, \cdots, g_{2n})$, then (1.9.7) can be rewritten as
$$y_0 = Ta.$$
Hence
$$a = T^{-1} y_0. \tag{1.9.12}$$
Let the third and fourth rows of T^{-1} be $\tilde{g}_3^T$ and $\tilde{g}_4^T$, then
$$a_3 = \tilde{g}_3^T y_0, \quad a_4 = \tilde{g}_4^T y_0.$$
By (1.1.13) we have
$$X^T = (T^{-1})^T \Lambda T^T,$$
that is
$$X^T (T^{-1})^T = (T^{-1})^T \Lambda.$$
Therefore $\tilde{g}_3$ and $\tilde{g}_4$ are the eigenvectors of the matrix X^T corresponding to the eigenvalues λ_3 and λ_4.

If the matrices X and K_z are obtained by the approach in §1.1 or §1.2, the matrix T is found during the procedure, so it is not hard to get a_3 and a_4 from (1.9.12). But if the matrices X and K_z are obtained by the approach in §1.3, then no eigenvalue problem has been solved beforehand, so we need to find $g_3, g_4, \tilde{g}_3$ and $\tilde{g}_4$. Now it is known that $\xi^{\frac{1}{2}}$ is close to λ_3 and λ_4, hence the inverse power method with the origin shift [47] is effective. After getting the desired two eigenvectors of X^T, we can use the above mentioned approach to make $\tilde{g}_3$ symmetric and $\tilde{g}_4$ antisymmetric with respect to the displacement. Besides we should choose an appropriate constant factor such that $\tilde{g}_3^T g_3 = 1$ and $\tilde{g}_4^T g_4 = 1$.

1.10 Exterior Stokes problems

The "Stokes paradox" is well known for the problems of stationary two dimensional viscous flow around an obstacle, hence this flow has no physical meaning. We consider three dimensional problems here. A simple one among the three dimensional problems is the axial symmetric flow, we consider the infinite element approximation of which in this section. The system of governing equations is

$$\mu \left(-\frac{\nabla(x \nabla u)}{x} + \frac{u}{x^2} \right) + \frac{\partial p}{\partial x} = 0, \tag{1.10.1}$$

$$-\mu \frac{\nabla(x \nabla v)}{x} + \frac{\partial p}{\partial y} = 0, \tag{1.10.2}$$

$$\frac{\partial}{\partial x}(xu) + \frac{\partial}{\partial y}(xv) = 0, \tag{1.10.3}$$

1.10 STOKES PROBLEMS

where the constant $\mu > 0$ is viscosity, (u, v) is the velocity field and p is the pressure field, and the y-axis is the axis of rotation. For the problem of external flow there is a boundary condition at the infinity, $(u, v) = (u_\infty, v_\infty)$. The constant vector $(-u_\infty, -v_\infty)$ added to the velocity field makes a new one $(u - u_\infty, v - v_\infty)$, hence there is no harm in assuming that the velocity vanishes at the infinity.

Generally speaking linear interpolation functions are not sufficient for the velocity field [63]. We apply triangular elements and take the vertices and middle points as nodes (Fig.16). Quadratic functions

$$a + bx + cy + dx^2 + exy + fy^2$$

are applied as the interpolation functions, then six coefficients are determined uniquely by the six nodal values. We assume that the pressure is a constant on each element.

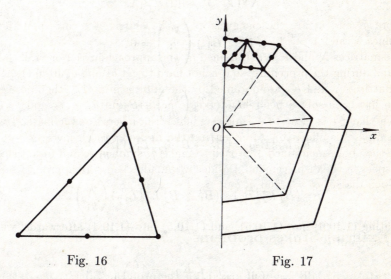

Fig. 16   Fig. 17

The partition is shown in Fig.17. We consider the k-th layer between Γ_{k-1} and Γ_k. If the number of triangular elements is N, then there are $N+1$ nodes on Γ_{k-1} (or Γ_k). Inside the k-th layer there are other $N+1$ nodes. The total number of nodes on one layer is $3(N+1)$. But we know by symmetry that $u = 0$ on the y-axis, so there are $2N$ degrees of freedom of the velocity components on Γ_{k-1} (or Γ_k), which are arranged as a $2N$ dimensional vector y_{k-1} (or y_k). There is an analogous $2N$ dimensional vector y^* associated with the interior nodes.

An arbitrary constant can be added to the pressure in the system (1.10.1)–(1.10.3). To remove this indefinite factor, we assign zero to the pressure on a fixed element in the k-th layer. Therefore the values of pressure in the k-th layer form a $N-1$ dimensional vector, which is denoted by q.

We evaluate the integral of "strain energy" in the first layer. By introducing some matrices, we have

$$W = \frac{1}{2}\mu \int x \left(|\nabla u|^2 + |\nabla v|^2 + \frac{u^2}{x^2} \right) dxdy$$

$$= \frac{1}{2} (y_0^T \quad y_1^T \quad y^{*T}) \begin{pmatrix} L_{11} & L_{12} \\ L_{12}^T & L_{22} \end{pmatrix} \begin{pmatrix} y_0 \\ y_1 \\ y^* \end{pmatrix}, \qquad (1.10.4)$$

$$-\int p \left(\frac{\partial}{\partial x}(xu) + \frac{\partial}{\partial y}(xv) \right) dxdy = (y_0^T \quad y_1^T \quad y^{*T}) \begin{pmatrix} B_1 \\ B_2 \end{pmatrix} q,$$

where L_{11} is a $4N \times 4N$ matrix, L_{22} is a $2N \times 2N$ matrix, B_1 is a $4N \times (N-1)$ matrix, and B_2 is a $2N \times (N-1)$ matrix.

The following system of equations

$$L_{12}^T \begin{pmatrix} y_0 \\ y_1 \end{pmatrix} + L_{22} y^* + B_2 q = 0,$$

$$(B_1^T \quad B_2^T) \begin{pmatrix} y_0 \\ y_1 \\ y^* \end{pmatrix} = 0$$

yields

$$y^* = -L_{22}^{-1}(L_{12}^T \begin{pmatrix} y_0 \\ y_1 \end{pmatrix} + B_2 q), \qquad (1.10.5)$$

$$B_2^T L_{22}^{-1} B_2 q = (B_1^T - B_2^T L_{22}^{-1} L_{12}^T) \begin{pmatrix} y_0 \\ y_1 \end{pmatrix}.$$

Thus

$$q = (B_2^T L_{22}^{-1} B_2)^{-1}(B_1^T - B_2^T L_{22}^{-1} L_{12}^T) \begin{pmatrix} y_0 \\ y_1 \end{pmatrix}. \qquad (1.10.6)$$

Substituting (1.10.6) into (1.10.5), and (1.10.5) into (1.10.4) afterwards, we obtain the expression of W,

$$W = \frac{1}{2} (y_0^T \quad y_1^T) \begin{pmatrix} K_0 & -A^T \\ -A & K_0' \end{pmatrix} \begin{pmatrix} y_0 \\ y_1 \end{pmatrix}$$

in terms of y_0 and y_1. The combined stiffness matrix associated with the k-th layer is

$$\xi^{k-1} \begin{pmatrix} K_0 & -A^T \\ -A & K_0' \end{pmatrix}.$$

The approach to calculate the combined stiffness matrix K_z on Ω is analogous to that in §1.5, but some new factors should be taken into account for the iterative method. We evaluate the flow

$$\zeta = \int_{\Gamma_0} x(u\cos(\nu, x) + v\cos(\nu, y)) \, ds$$

on Γ_0, where ν stands for the normal direction on Γ_0 pointing to the infinity. ζ depends on y_0 linearly, hence we can get a $2N$ dimensional column vector h, such that
$$\zeta = h^T y_0.$$
It is easy to see that the flow on Γ_k is $\zeta = \xi^{2k} h^T y_k$.

Let the combined stiffness matrix of the 2^l layers between Γ_0 and Γ_{2^l} be
$$\begin{pmatrix} K_l & -A_l^T \\ -A_l & K_l' \end{pmatrix}.$$
In the process of calculating it, pressure p vanishes on a certain element. For definiteness, we assume that this element is assigned on the 2^l-th layer.

Let $k = 2^l$, then the "strain energy" on $2k$ layers is
$$\begin{aligned} W = &\frac{1}{2}(y_0^T \ y_k^T) \begin{pmatrix} K_l & -A_l^T \\ -A_l & K_l' \end{pmatrix} \begin{pmatrix} y_0 \\ y_k \end{pmatrix} \\ &+ \frac{1}{2}\xi^k (y_k^T \ y_{2k}^T) \begin{pmatrix} K_l & -A_l^T \\ -A_l & K_l' \end{pmatrix} \begin{pmatrix} y_k \\ y_{2k} \end{pmatrix}. \end{aligned} \tag{1.10.7}$$

It should be noticed that p does not vanish on two elements but only one. Now that p vanishes on an element of the $2k$-th layer, so p should be arbitrary constant on each element of the first k layers. Intergrating the equation (1.10.3) on these k layers we obtain
$$h^T y_0 = \xi^{2k} h^T y_k. \tag{1.10.8}$$
Therefore y_k is not arbitrary in the procedure of minimizing W.

We introduce a Lagrangian multiplier η and set
$$L = W + \eta h^T (\xi^{2k} y_k - y_0).$$
Letting $\frac{\partial L}{\partial y_k} = 0, \frac{\partial L}{\partial \eta} = 0$, we obtain (1.10.8) and
$$-A_l y_0 + K_l' y_k + \xi^k(-A_l^T y_{2k} + K_l y_k) + \eta \xi^{2k} h = 0,$$
hence
$$y_k = (\xi^k K_l + K_l')^{-1}(A_l y_0 + \xi^k A_l^T y_{2k} - \eta \xi^{2k} h). \tag{1.10.9}$$
Substituting (1.10.9) into (1.10.8) we obtain
$$h^T y_0 = \xi^{2k} h^T (\xi^k K_l + K_l')^{-1}(A_l y_0 + \xi^k A_l^T y_{2k} - \eta \xi^{2k} h).$$
Then we can solve
$$\eta = \frac{\xi^{-4k}}{h^T (\xi^k K_l + K_l')^{-1} h} h^T \{\xi^{2k}(\xi^k K_l + K_l')^{-1}(A_l y_0 + \xi^k A_l^T y_{2k}) - y_0\},$$

which is abbreviated to
$$\eta = h_k^T y_0 + h_{2k}^T y_{2k},$$

where h_k and h_{2k} are $2N$-dimensional column vectors. We substitute this expression into (1.10.9) and get

$$y_k = (\xi^k K_l + K_l')^{-1}\{A_l y_0 + \xi^k A_l^T y_{2k} - \xi^{2k} h(h_k^T y_0 + h_{2k}^T y_{2k})\}. \tag{1.10.10}$$

Then substituting it into (1.10.7) we obtain

$$W = \frac{1}{2}(y_0^T \; y_{2k}^T)\begin{pmatrix} K_{l+1} & -A_{l+1}^T \\ -A_{l+1} & K_{k+1}' \end{pmatrix}\begin{pmatrix} y_0 \\ y_{2k} \end{pmatrix}$$

after some calculation, where

$$K_{l+1} = K_l - (A_l^T + \xi^{2k} h_k h^T)(\xi^k K_l + K_l')^{-1}(A_l - \xi^{2k} h h_k^T), \tag{1.10.11}$$

$$K_{l+1}' = \xi^k K_l' - (\xi^k A_l + \xi^{2k} h_{2k} h^T)(\xi^k K_l + K_l')^{-1}(\xi^k A_l^T - \xi^{2k} h h_{2k}^T), \tag{1.10.12}$$

$$A_{l+1} = \xi^k A_l (\xi^k K_l + K_l')^{-1} A_l - \xi^{4k} h_{2k} h^T (\xi^k K_l + K_l')^{-1} h h_k^T. \tag{1.10.13}$$

Now we can evaluate the combined stiffness matrices for 2^l layers, $l = 0, 1, 2, \cdots$, by (1.10.11)–(1.10.13) successively beginning from $l = 0$. Suppose K_l, K_l' and A_l are known for one natrual number l.

We take an arbitrary vector $g \in \mathbb{R}^{2N}$ such that $h^T g \neq 0$. For instance we may take $g = \varepsilon_1$, where $\varepsilon_1 = (1, 0, \cdots, 0)^T$. Set $y_{2^l} = \alpha \varepsilon_1$, where α is a constant. Then by (1.10.8) we obtain

$$h^T y_0 = \alpha \xi^{2^{l+1}} h^T \varepsilon_1, \tag{1.10.14}$$

that is
$$\alpha = \xi^{-2^{l+1}} \frac{h^T y_0}{h^T \varepsilon_1}.$$

Now we have

$$W = \frac{1}{2}(y_0^T \; \alpha \varepsilon_1^T)\begin{pmatrix} K_l & -A_l^T \\ -A_l & K_l' \end{pmatrix}\begin{pmatrix} y_0 \\ \alpha \varepsilon_1 \end{pmatrix}$$

$$= \frac{1}{2}\left(y_0^T \; \frac{\xi^{-2^{l+1}}}{h^T \varepsilon_1} y_0^T h \varepsilon_1^T\right)\begin{pmatrix} K_l & -A_l^T \\ -A_l & K_l' \end{pmatrix}\begin{pmatrix} y_0 \\ \frac{\xi^{-2^{l+1}}}{h^T \varepsilon_1} \varepsilon_1 h^T y_0 \end{pmatrix}.$$

It can be arranged as
$$W = \frac{1}{2} y_0^T K_z^{(l)} y_0,$$

where
$$K_z^{(l)} = K_l - \frac{\xi^{-2^{l+1}}}{h^T \varepsilon_1}(h \varepsilon_1^T A_l + A_l^T \varepsilon_1 h^T) + \frac{\xi^{-2^{l+2}}}{(h^T \varepsilon_1)^2} h \varepsilon_1^T K_l' \varepsilon_1 h^T.$$

1.10 STOKES PROBLEMS

We will prove in the next chapter that
$$\lim_{l\to\infty} K_z^{(l)} = K_z.$$

By analogy with §1.5, we can derive an analogous formula for the iterative method of the second type,

$$y_0^T K_z y_0 = \min_w \left\{ (y_0^T \ w^T) \begin{pmatrix} K_l & -A_l^T \\ -A_l & K_l' \end{pmatrix} \begin{pmatrix} y_0 \\ w \end{pmatrix} + \xi^{2^l} w^T K_z w \right\}. \tag{1.10.15}$$

But now w is not arbitrary, and from (1.10.8) it is restricted by

$$h^T y_0 = \xi^{2^{l+1}} h^T w. \tag{1.10.16}$$

We introduce a Lagrangian multiplier η and set

$$L = (y_0^T \ w^T) \begin{pmatrix} K_l & -A_l^T \\ -A_l & K_l' \end{pmatrix} \begin{pmatrix} y_0 \\ w \end{pmatrix} + \xi^{2^l} w^T K_z w + \eta h^T (\xi^{2^{l+1}} w - y_0).$$

Letting $\frac{\partial L}{\partial w} = 0, \frac{\partial L}{\partial \eta} = 0$, we obtain (1.10.16) and

$$-2A_l y_0 + 2K_l' w + 2\xi^{2^l} K_z w + \eta \xi^{2^{l+1}} h = 0,$$

hence

$$w = (\xi^{2^l} K_z + K_l')^{-1} \left(A_l y_0 - \frac{1}{2} \eta \xi^{2^{l+1}} h \right). \tag{1.10.17}$$

Substituting it into (1.10.16) we obtain

$$h^T y_0 = \xi^{2^{l+1}} h^T (\xi^{2^l} K_z + K_l')^{-1} \left(A_l y_0 - \frac{1}{2} \eta \xi^{2^{l+1}} h \right).$$

Then we can solve

$$\eta = \frac{2\xi^{-2^{l+2}}}{h^T (\xi^{2^l} K_z + K_l')^{-1} h} h^T \{ \xi^{2^{l+1}} (\xi^{2^l} K_z + K_l')^{-1} A_l y_0 - y_0 \},$$

which is abbreviated to
$$\eta = 2 r_l^T y_0,$$

where r_l is a $2N$ dimensional column vector. We substitute this expression into (1.10.17) and get

$$w = (\xi^{2^l} K_z + K_l')^{-1} (A_l y_0 - \xi^{2^{l+1}} h r_l^T y_0). \tag{1.10.18}$$

Then substituting it into (1.10.15), we obtain

$$K_z = K_l - (A_l^T - \xi^{2^{l+1}} r_l h^T)(\xi^{2^l} K_z + K_l')^{-1} (A_l + \xi^{2^{l+1}} h r_l^T)$$

after some calculation. Then we have the iterative scheme

$$K_z^{(m+1)} = K_l - (A_l^T - \xi^{2^{l+1}} r_l h^T)(\xi^{2^l} K_z^{(m)} + K_l')^{-1}(A_l + \xi^{2^{l+1}} h r_l^T).$$

We will prove in the next chapter that if only $K_z^{(0)}$ is a symmetric and positive definite matrix, then

$$\lim_{m \to \infty} K_z^{(m)} = K_z.$$

We turn now to consider the calculation of the transfer matrix X. Taking $l = 0$ and $k = 1$ in (1.10.10), we have an equation,

$$-(A - \xi^2 h h_1^T) y_0 + (\xi K_0 + K_0') y_1 - (\xi A^T - \xi^2 h h_2^T) y_2 = 0.$$

The relation

$$y_k = X^k y_0, \quad k = 1, 2, \cdots,$$

is substituted into it, then we get the equation for X,

$$(\xi A^T - \xi^2 h h_2^T) X^2 - (\xi K_0 + K_0') X + (A - \xi^2 h h_1^T) = 0.$$

Noting that all eigenvalues λ satisfy $|\lambda| < \xi^{-\frac{1}{2}}$, we can evaluate the matrix X by means of the approach in §1.1. It is simple to evaluate X if K_z is known. Letting $l = 0$ in (1.10.18) we get

$$y_1 = (\xi K_z + K_0')^{-1}(A - \xi^2 h r_0^T) y_0,$$

hence

$$X = (\xi K_z + K_0')^{-1}(A - \xi^2 h r_0^T).$$

The link between the combined elements and other conventional elements is as follows: If the conventional elements occupy a domain Ω^*, and if a vector z consists of the components of the nodal velocity values on Ω^*, and a vector q consists of the pressure values on elements, then we define

$$\frac{1}{2}\int_{\Omega^*} x\left(|\nabla u|^2 + |\nabla v|^2 + \frac{u^2}{x^2}\right) dxdy = \frac{1}{2} z^T K^* z,$$

$$-\int_{\Omega^*} p\left(\frac{\partial}{\partial x}(xu) + \frac{\partial}{\partial y}(xv)\right) dxdy = z^T B^* q.$$

Let the union of domains Ω and Ω^* be Ω_0. The velocity is known on the boundary of Ω_0. Among the components of the vector y_0, some of them coincide with that of z, some of them are the boundary velocity, and some components of z are also boundary velocity. Assume that a vector z_1 consists of the interior velocity components of z. We set

$$W = \frac{1}{2} y_0^T K_z y_0 + \frac{1}{2} z^T K^* z + z^T B^* q,$$

then let $\frac{\partial W}{\partial z_1} = 0$ and $\frac{\partial W}{\partial q} = 0$, after those procedure an algebriac system of equations for z_1 and q is obtained, then we can solve it.

We will verify the above algorithm in the next chapter.

1.11 Nonhomogeneous equations and nonhomogeneous boundary conditions

The equations which we have considered in the preceding sections are all homogeneous. As for the boundary conditions, at least on the part of boundary which we are concerned about, they are homogeneous too. In this section we consider the infinite element method for nonhomogeneous problems. We take the corner problem with the equation,
$$-\triangle u = p,$$
and the boundary condition near the point O,
$$u = f$$
as an example. Because the neighborhood Ω can be taken appropriately small, we may apply some simple functions to be the approximations of p and f. Therefore it is reasonable to assume that p is a constant in Ω, and f is linear on both adjacent sides Γ^* and Γ_* to the point O.

We are going to construct a particular solution to the equation and the boundary condition. We define a linear function $u_1(x,y) = ax + by + c$ such that it satisfies the boundary condition along the boundary. Clearly we have $\triangle u_1 = 0$. Next we define a function u_2 such that $-\triangle u_2 = p$ and u_2 vanishes on Γ^* and Γ_*. Let the equations of Γ^*, Γ_* be
$$A_i x + B_i y + C_i = 0, \quad i = 1, 2,$$
then we set
$$u_2 = D(A_1 x + B_1 y + C_1)(A_2 x + B_2 y + C_2),$$
thus
$$-\triangle u_2 = -2D(A_1 A_2 + B_1 B_2).$$
Set
$$D = -\frac{p}{2(A_1 A_2 + B_1 B_2)},$$
then u_2 is the desired one. If the two line segments are perpendicular to each other, then $A_1 A_2 + B_1 B_2 = 0$, it is impossible to construct a quadratic function as the particular solution. We may use cubic functions.

It is more troublesome to obtain a particular solution for general boundary value f and general nonhomogeneous term p. It is easy to find a function satisfying the boundary condition. Let it be u_1. We continue to construct a function u_2, such that
$$-\triangle u_2 = p + \triangle u_1,$$

and u_2 vanishes on Γ^* and Γ_*, then $u_1 + u_2$ is the desired particular solution.

The solution u_2 can be expressed in terms of series. We introduce an infinite algebraic system

$$-Ay_{k-1} + Ky_k - A^T y_{k+1} = p_k, \quad k = 1, 2, \cdots, \tag{1.11.1}$$

analogous to (1.1.6), where p_k is the "load vector" associated with $p + \triangle u_1$. Suppose the transfer matrix X has already been found, and we extend Γ^* and Γ_* to the infinite, then an unbounded domain outside Γ_0 is formed. There is another transfer matrix on this unbounded domain (see §1.6). To make a distinction between them, the later one is denoted by $\tilde{X}$. Then we solve the following series of algebraic systems

$$(K - A\tilde{X} - A^T X) z_m^{(m)} = p_m, \quad m = 1, 2, \cdots. \tag{1.11.2}$$

The matrices of the coefficients of systems (1.11.2) keep fixed for all m, hence the amount of work to solve (1.11.2) is not too much. Let

$$z_{k+1}^{(m)} = X z_k^{(m)}, \quad \text{for } k \geq m,$$
$$z_{k-1}^{(m)} = \tilde{X} z_k^{(m)}, \quad \text{for } k \leq m.$$

Summing them up with respect to m, we get

$$z_k = \sum_{k=1}^{\infty} z_k^{(m)}, \quad k = 0, 1, \cdots.$$

The interpolation function of z_k is u_2.

If the Neumann boundary condition $\frac{\partial u}{\partial \nu} = 0$ is considered, we set

$$p_0 = -\sum_{k=1}^{\infty} p_k,$$

then solve the following series of algebraic systems ($m = 1, 2, \cdots$)

$$\begin{cases} (K - A\tilde{X}) z_{m-1}^{(m)} - A^T z_m^{(m)} = \tilde{p}_{m-1}, \\ -A z_{m-1}^{(m)} + (K - A^T X) z_m^{(m)} = -\tilde{p}_{m-1}, \end{cases} \tag{1.11.3}$$

where

$$\tilde{p}_0 = p_0,$$
$$\tilde{p}_m = \tilde{p}_{m-1} + p_m, \quad m = 1, 2, \cdots.$$

The matrix of the coefficients of system (1.11.3) is singular. We may set the last component of $z_m^{(m)}$ to be zero, and cross out the corresponding row and column of

1.11 NONHOMOGENEOUS EQUATIONS

the matrix of the coefficients. After such treatment, the system (1.11.3) is solvable. We set

$$z_{k+1}^{(m)} = X z_k^{(m)}, \quad \text{for } k \geq m,$$
$$z_{k-1}^{(m)} = \tilde{X} z_k^{(m)}, \quad \text{for } k \leq m-1,$$

and sum them up with respect to m, then we obtain u_2.

We will verify the above algorithm in the next chapter.

In brief, we have already found a particular solution u_0 on Ω to the equation and the local boundary condition. Let $u = u_0 + v$, then v satisfies a homogeneous equation and a homogeneous boundary condition on Ω. It can be solved by means of the approach in §1.7, and the conventional finite element method is applied to solve u outside Ω. Some conditions on Γ_0 for u and v are necessary for solving them simultaneously. Let us derive these conditions.

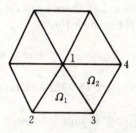

Fig. 18

See Fig.18, if b_1 is a node on Γ_0, and surrounding it are the nodes $b_2, b_3, \cdots, b_{m+1}$, and the adjacent elements are $e_1, e_2, \cdots, e_m$, where $e_1, \cdots, e_{m_1}$ lie inside Ω and $e_{m_1+1}, \cdots, e_m$ outside Ω. Function φ_1 is equal to 1 at the point b_1 and vanishes at other nodes, and is linear on each element. We construct equations at the point b_1 as follows:

$$u(b_1) = u_0(b_1) + v(b_1), \tag{1.11.4}$$

$$\sum_{i=1}^{m_1} \int_{\Omega_i} \nabla\varphi_1 \cdot \nabla u_0 \, dxdy + \sum_{i=1}^{m_1} \int_{\Omega_i} \nabla\varphi_1 \cdot \nabla v \, dxdy$$
$$+ \sum_{i=m_1+1}^{m} \int_{\Omega_i} \nabla\varphi_1 \cdot \nabla u \, dxdy = \sum_{i=1}^{m} \int_{\Omega_i} \varphi_1 p \, dxdy, \tag{1.11.5}$$

where $\nabla\varphi_1 = (\frac{\partial \varphi_1}{\partial x}, \frac{\partial \varphi_1}{\partial y})$, etc. The second and the third integral at the left hand side of (1.11.5) can be expressed by the stiffness matrices and nodal values.

We can solve simultaneous equations involving (1.11.4),(1.11.5), the conventional finite element equations outside Ω, and the equation

$$y_1 = X y_0,$$

and get the desired solution.

We conclude this section with discontinuous boundary value problems. Generally speaking the strain energy is infinite at this case, therefore the stiffness matrix in normal sense fails to exist. The use of the infinite element method can overcome this difficulty satisfactorily. We consider the Laplace equation and assume that the solution near a corner is equal to a constant α on Γ^* and a constant β on Γ_* shown in Fig. 9, with $\alpha \neq \beta$.

As usual we take the infinite element partition as §1.7. The nodal values on the boundaries Γ^*, Γ_* are not included in y_k. Using the total stiffness matrix we express the strain energy on one layer as:

$$W = \frac{1}{2}(y_{k-1}^T \quad \alpha \quad \beta \quad y_k^T) \begin{pmatrix} L_{11} & L_{12} & L_{13} & L_{14} \\ L_{21} & L_{22} & L_{23} & L_{24} \\ L_{31} & L_{32} & L_{33} & L_{34} \\ L_{41} & L_{42} & L_{43} & L_{44} \end{pmatrix} \begin{pmatrix} y_{k-1} \\ \alpha \\ \beta \\ y_k \end{pmatrix} \quad (1.11.6)$$

We will prove in the next chapter that there exist a $(n-2) \times (n-2)$ real matrix X and a $(n-2)$-dimensional vector g, such that

$$y_{k+1} = Xy_k + g, \qquad k = 0, 1, \cdots. \quad (1.11.7)$$

Applying (1.11.6) we can write down the finite element equation on Γ_1 as:

$$L_{41}y_0 + L_{42}\alpha + L_{43}\beta + (L_{44} + L_{11})y_1 + L_{12}\alpha + L_{13}\beta + L_{14}y_2 = 0. \quad (1.11.8)$$

We substitute (1.11.7) into (1.11.8) and obtain

$$(L_{41} + (L_{44}+L_{11})X + L_{14}X^2)y_0 + (L_{12} + L_{42})\alpha$$
$$+ (L_{13} + L_{43})\beta + (L_{44} + L_{11})g + L_{14}(X + I)g = 0.$$

Since y_0 is arbitrary, we set it to be zero, and obtain

$$g = -(L_{44} + L_{11} + L_{14}X + L_{14})^{-1}((L_{12} + L_{42})\alpha + (L_{13} + L_{43})\beta). \quad (1.11.9)$$

Again we note y_0 is arbitrary and get

$$A^T X^2 - KX + A = 0, \quad (1.11.10)$$

where $K = L_{44} + L_{11}, A = -L_{41}$, thus X is just the transfer matrix under homogeneous boundary conditions. The exterior part to Ω of the domain is divided into finite elements in a conventional way. To solve equations simultaneously, we may use some relations analogous to (1.11.4),(1.11.5), or we may evaluate the load vector on Γ_0. For the later case we substitute (1.11.7) into equation

$$L_{11}y_0 + L_{12}\alpha + L_{13}\beta + L_{14}y_1 = f_0,$$

and obtain by (1.11.9) that

$$f_0 = K_z y_0 + E\alpha + F\beta, \qquad (1.11.11)$$

where K_z is the combined stiffness matrix under homogeneous boundary condition, and

$$E = L_{12} - L_{14}(L_{44} + L_{11} + L_{14}X + L_{14})^{-1}(L_{12} + L_{42}),$$
$$F = L_{13} - L_{14}(L_{44} + L_{11} + L_{14}X + L_{14})^{-1}(L_{13} + L_{43}).$$

We evaluate K_z, E, F first, then use (1.11.11) to assemble the combined element and other conventional finite elements, then solve the coupled equations.

1.12 Boundary value problems of the biharmonic equation

Let us consider corner problems here. The biharmonic equation is

$$\triangle^2 u = 0. \qquad (1.12.1)$$

We assume Dirichlet boundary conditions,

$$u = f_1, \qquad \frac{\partial u}{\partial \nu} = f_2$$

on the boundary. At first we suppose f_1 and f_2 vanish on the adjacent sides Γ^* and Γ_* to the point O (Fig.9).

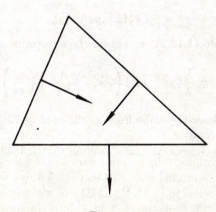

Fig. 19

The infinite element mesh is shown in Fig.10. For definiteness we use Morley triangular elements here [61]. The values of u are given at the vertices and the normal derivatives of u are given at the middle points (Fig.19). Quadratic functions are applied as the interpolation functions and six coefficients are determined uniquely by the six nodal values. Here it should be noticed that at each middle point we must fix one normal direction and all layers must be similar, including the direction.

The strain energy associated with (1.12.1) is

$$W = \frac{1}{2}\int_\Omega \left\{ \left(\frac{\partial^2 u}{\partial x^2} + \sigma \frac{\partial^2 u}{\partial y^2}\right)\frac{\partial^2 u}{\partial x^2} + 2(1-\sigma)\left(\frac{\partial^2 u}{\partial x \partial y}\right)^2 \right. $$
$$\left. + \left(\sigma\frac{\partial^2 u}{\partial x^2} + \frac{\partial^2 u}{\partial y^2}\right)\frac{\partial^2 u}{\partial y^2} \right\} dxdy, \tag{1.12.2}$$

where the constant $\sigma \in [0, \frac{1}{2})$. We consider the k-th layer between Γ_{k-1} and Γ_k. If the number of triangular elements is N, then there are $N-1$ nodes on Γ_{k-1} (or Γ_k) and inside the k-th layer there are other $N-1$ nodes, where the nodes on Γ^* and Γ_* are not taken into account. Let y_{k-1} (or y_k) be a vector with components of $\frac{u}{r}$ or $\frac{\partial u}{\partial \nu}$ at the nodes on Γ_{k-1} (or Γ_k), and y_k^* be a vector with components of $\frac{\partial u}{\partial \nu}$ at the interior nodes, where r stands for the distance between a point to the point O. It is easy to see by (1.12.2) that the total stiffness matrices of all layers are just the same. Upon introducing some matrices, we have the strain energy on the k-th layer:

$$W = \frac{1}{2}(y_0^T \ y_1^{*T} \ y_1^T) \begin{pmatrix} L_{11} & L_{12} & L_{13} \\ L_{12}^T & L_{22} & L_{23} \\ L_{13}^T & L_{23}^T & L_{33} \end{pmatrix} \begin{pmatrix} y_0 \\ y_1^* \\ y_1 \end{pmatrix}. \tag{1.12.3}$$

From the equation

$$L_{12}^T y_0 + L_{22} y_1^* + L_{23} y_1 = 0,$$

we obtain

$$y_1^* = -L_{22}^{-1}(L_{12}^T y_0 + L_{23} y_1). \tag{1.12.4}$$

Substituting (1.12.4) into (1.12.3), we obtain the expression of W,

$$W = \frac{1}{2}(y_0^T \ y_1^T)\begin{pmatrix} K_0 & -A^T \\ -A & K_0' \end{pmatrix}\begin{pmatrix} y_0 \\ y_1 \end{pmatrix}.$$

Then the argument follows the same lines as those of §1.7. The eigenvalues of the transfer matrix X satisfy $|\lambda| < 1$.

Let us consider a more general case. Assume that u is still zero on Γ^* and Γ_*, but

$$\left.\frac{\partial u}{\partial \nu}\right|_{\Gamma^*} = \alpha, \quad \left.\frac{\partial u}{\partial \nu}\right|_{\Gamma_*} = \beta,$$

where α and β are constants. Following §1.11, we derive the expressions of y_{k+1} and f_0. Different from the homogeneous case, there are two more degrees of freedom in each layer. Therefore we have a 5×5 block matrix $L = (L_{ij})_{i,j=1}^{5}$ and the strain energy in the k-th layer is

$$W = \frac{1}{2} (y_{k-1}^T \quad \alpha \quad y_k^{*T} \quad \beta \quad y_k) L \begin{pmatrix} y_{k-1} \\ \alpha \\ y_k^* \\ \beta \\ y_k \end{pmatrix},$$

where the meaning of y_{k-1}, y_k^*, y_k is the same as before. Upon assembling the elements in the first and the second layers, and constructing the equation equation with respect to Γ_1, we have

$$L_{51}y_0 + L_{52}\alpha + L_{53}y_1^* + L_{54}\beta + (L_{55}+L_{11})y_1 + L_{12}\alpha + L_{13}y_2^* + L_{14}\beta + L_{15}y_2 = 0. \tag{1.12.5}$$

Taking the nodes inside the layers into account, we get the equations

$$L_{31}y_0 + L_{32}\alpha + L_{33}y_1^* + L_{34}\beta + L_{35}y_1 = 0, \tag{1.12.6}$$

$$L_{31}y_1 + L_{32}\alpha + L_{33}y_2^* + L_{34}\beta + L_{35}y_2 = 0. \tag{1.12.7}$$

We substitute (1.12.6),(1.12.7) into (1.12.5) and eliminate y_1^*, y_2^*, then we get an equation as

$$-Ay_0 + Ky_1 - A^T y_2 = f, \tag{1.12.8}$$

where

$$A = L_{53}L_{33}^{-1}L_{31} - L_{51},$$
$$K = L_{55} + L_{11} - L_{53}L_{33}^{-1}L_{35} - L_{13}L_{33}^{-1}L_{31},$$
$$f = (-L_{52} - L_{12} + L_{53}L_{33}^{-1}L_{32} + L_{13}L_{33}^{-1}L_{32})\alpha$$
$$+ (-L_{54} - L_{14} + L_{53}L_{33}^{-1}L_{34} + L_{13}L_{33}^{-1}L_{34})\beta.$$

We have the same relation as (1.11.7),

$$y_{k+1} = Xy_k + g, k = 0, 1, \cdots. \tag{1.12.9}$$

Substituting (1.12.9) into (1.12.8) we obtain

$$-Ay_0 + KXy_0 - A^T X^2 y_0 + Kg - A^T(Xg + g) = f,$$

then we have

$$A^T X^2 - KX + A = 0,$$

and

$$g = (K - A^T X - A^T)^{-1} f.$$

So we can solve this problem in the same way as §1.11.

1.13 Multigrid algorithm

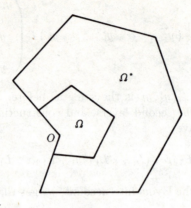

Fig. 20

This section is devoted to the following problem. A polygonal domain Ω_0 is divided into two parts, Ω and Ω^* (Fig. 20), and a function u is the solution to the following problem on Ω_0,

$$-\triangle u = f,$$

$$u\big|_{\partial\Omega_0} = 0,$$

where $\partial\Omega_0$ is the boundary of Ω_0. For the sake of simplicity, we assume that f vanishes on Ω. Ω_0 is partitioned into triangular elements, where the local partition on Ω is an infinite similar one, and the local partition on Ω^* is a conventional finite element one. So far we have obtained the coarse mesh, denoted by T_1.

To get the fine mesh T_2, we take $\xi^{\frac{k}{2}}$ $(k = 1, 2, \ldots)$ to be the constants of proportionality and make an infinite similar partition on Ω. We require that the new mesh is a refinement of T_1. Such a partition is unique and each line segment is divided into two parts with ratio $1 : \xi^{\frac{1}{2}}$. At the same time each element in Ω^* is divided into four triangles appropriately. By this way we can get further fine meshes $T_j, j = 2, \ldots, J$.

Let us apply linear interpolation for nodal values. The combined stiffness matrix on Ω for T_j is denoted by K_z^j and the transfer matrix is denoted by X_j. In the multigrid iterative procedure, the domain Ω is regarded as a single element. Then we have a series of algebraic systems with finite order,

$$A_j u_j = f_j, \qquad j = 1, 2, \cdots, J, \tag{1.13.1}$$

and u_J is the desired solution.

The iterative procedure is normal. For any vector u_{j-1} defined on the nodes of T_{j-1}, by the matrix X_{j-1} and linear interpolation we get a function defined on the closed domain $\bar{\Omega}_0$. Then those values on the nodes of $\mathcal{T}_j$ are arranged as a vector u_j. The prolongation matrix I_j is defined as

$$u_j = I_j u_{j-1}.$$

The restriction matrix is defined as

$$I_j^t = \frac{1}{4} I_j^T.$$

Let a matrix R_j represent one iterative scheme for solving (1.13.1), that is, the equation

$$u_j^{(m+1)} = u_j^{(m)} - R_j(A_j u_j^{(m)} - f_j)$$

creates a series of vectors, $u_j^{(0)}, u_j^{(1)}, \ldots$, which converges to u_j. For simplicity we assume that R_j is symmetric. Then the multigrid iterative scheme is

$$u_J^{(m+1)} = u_J^{(m)} - B_J(A_J u_J^{(m)} - f_J), \quad m = 0, 1, \ldots,$$

where the matrices $B_1, B_2, \ldots, B_J$ is defined by recurrence:
(1) $B_1 = A_1^{-1}$.
(2) If B_{j-1} has been defined, then for any vector g on the nodes of $\mathcal{T}_j$, we define $B_j g$ as follows:
(a) Pre-smoothing:
$$w^0 = 0,$$
$$w^l = w^{l-1} + R_j(g - A_j w^{l-1}), \quad l = 1, 2, \ldots, m_j.$$
(b) Correction:
$$w^{m_j+1} = w^{m_j} + I_j q^p,$$
where
$$q^0 = 0,$$
$$q^l = (I - B_{j-1} A_{j-1}) q^{l-1} + B_{j-1} I_j^t (g - A_j w^{m_j}), \quad l = 1, \ldots, p.$$
(c) Post-smoothing:
$$w^l = w^{l-1} + R_j(g - A_j w^{l-1}), \quad l = m_j + 2, \ldots, 2m_j + 1.$$
(d) Finally, set
$$B_j g = w^{2m_j+1}.$$

If $p = 1, m_j \equiv m$, it is a V-circle, if $p = 2, m_j \equiv m$, it is a W-circle, and if $p = 1, \gamma_0 m_j \leq m_{j-1} \leq \gamma_1 m_j, \gamma_0 > 1$, it is a generalized V-circle.
We will study the convergence problem in Chapter Three.

1.14 Helmholtz equation

We consider the infinite element method for the Helmholtz equation

$$-\triangle u + \lambda u = 0, \tag{1.14.1}$$

where λ is a complex number. Since the stiffness matrices on the layers change with the level, the approaches in the previous sections are not applicable to this equation.

For example we consider the corner problem. Let the boundary condition be $\frac{\partial u}{\partial \nu} = 0$ on Γ^* and Γ_* of Fig.9. We partition the domain as Fig.10. The stiffness matrix for the k-th layer is

$$\begin{pmatrix} K_0 & -A^T \\ -A & K_0' \end{pmatrix} + \lambda \xi^{2(k-1)} \begin{pmatrix} L_0 & -D^T \\ -D & L_0' \end{pmatrix},$$

where the matrices K_0, K_0' and A are the same as that of §1.7, the second matrix is a symmetric and positive definite one. We get a system of equations analogous to (1.1.6),

$$(K_0 + \lambda L_0)y_0 - (A^T + \lambda D^T)y_1 = f_0, \tag{1.14.2}_0$$

$$-(A + \lambda D)y_0 + (K + \lambda L)y_1 - (A^T + \lambda \xi^2 D^T)y_2 = 0, \tag{1.14.2}_1$$

$$\cdots\cdots\cdots\cdots$$

$$-(A + \lambda \xi^{2(k-1)}D)y_{k-1} + (K + \lambda \xi^{2(k-1)}L)y_k$$
$$- (A^T + \lambda \xi^{2k} D^T)y_{k+1} = 0, \tag{1.14.2}_k$$

$$\cdots\cdots\cdots\cdots$$

where $K = K_0' + K_0$, and $L = L_0' + \xi^2 L_0$. If λ is not an eigenvalue of the equation (1.14.1) on the domain Ω, then there exists a matrix $X(\lambda)$, such that

$$y_1 = X(\lambda)y_0.$$

And if λ is a real number, then $X(\lambda)$ is a real matrix. Making a similarity transformation $x \to \xi^{-k}x, y \to \xi^{-k}y$, then Γ_k is transferred into Γ_0, and the equation (1.14.1) becomes

$$-\triangle u + \lambda \xi^{2k} u = 0. \tag{1.14.3}$$

Therefore if $\lambda \xi^{2k}$ is not an eigenvalue on the domain Ω, then

$$y_{k+1} = X(\lambda \xi^{2k})y_k.$$

If neither λ nor $\lambda \xi^2$ is an eigenvalue, by $(1.14.2)_1$ we have

$$-(A + \lambda D) + (K + \lambda L)X(\lambda) - (A^T + \lambda \xi^2 D^T)X(\lambda \xi^2)X(\lambda) = 0. \tag{1.14.4}$$

1.14 HELMHOLTZ EQUATION

We will prove in the next chapter that $X(\lambda)$ is an analytical function with the independent variable λ, hence $X(\lambda)$ can be developed into a power series in the neighborhood of the point $\lambda = 0$,

$$X(\lambda) = X_0 + \lambda X_1 + \cdots + \lambda^m X_m + \cdots, \tag{1.14.5}$$

where $X_0 = X(0)$, which is the very transfer matrix X obtained in §1.7. We substitute (1.14.5) into (1.14.4), then arrange the terms according to the power of λ, consequently we obtain a system

$$-A + KX_0 - A^T X_0^2 = 0,$$
$$D + KX_1 + LX_0 - A^T(X_0 X_1 + \xi^2 X_1 X_0) - \xi^2 D^T X_0^2 = 0,$$
$$\cdots\cdots\cdots\cdots$$

in general

$$KX_m + LX_{m-1} - A^T \sum_{i=0}^{m} \xi^{2i} X_i X_{m-i} - \xi^2 D^T \sum_{i=0}^{m-1} \xi^{2i} X_i X_{m-i-1} = 0,$$

$$m = 2, 3, \cdots$$

where the first one has been solved in §1.7. Beginning from the second one the equations are linear and the unknowns are $X_1, X_2, \cdots$ respectively. We put it in a nutshell, then they are in the form of

$$J_m X_m + A_m^T X_m M_m = F_m.$$

If we multiply it by $(A_m^T)^{-1}$ on the left, then we get the equation in a canonical form

$$(A_m^T)^{-1} J_m X_m + X_m M_m = (A_m^T)^{-1} F_m.$$

Solving it we get X_m.

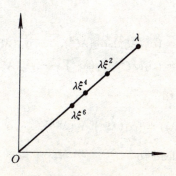

Fig. 21

(1.14.5) is the expression of $X(\lambda)$ in a neighborhood of the point O. For such λ with a greater absolute value we can find $X(\lambda)$ from $X(\lambda\xi^2)$ by the equation (1.14.4). $X(\lambda\xi^{2k})$ can be found from (1.14.5) for sufficiently large k provided $\lambda, \lambda\xi^2, \cdots, \lambda\xi^{2k}$ are not eigenvalues, then (1.14.4) is applied as a recurrence relation.

Let
$$K_z(\lambda) = (K_0 + \lambda L_0) - (A^T + \lambda D^T)X(\lambda),$$
then it is the combined stiffness matrix on the domain Ω.

If we need to evaluate the singularity of the solution at the point O, then the solution can be expressed as
$$y_k \sim \alpha(\lambda)g_1 + \beta(\lambda)g_2\lambda_2^k, \tag{1.14.6}$$
where $g_1 = (1, 1, \cdots, 1)^T$ corresponds to the eigenvalue $\lambda_1 = 1$, and λ_2, g_2 are a couple of eigenvalue and eigenvector representing the singularity of the solution. $\alpha(\lambda)$ and $\beta(\lambda)$ are determined by y_0, hence there are analytical functions $\alpha_1(\lambda), \alpha_2(\lambda), \cdots, \alpha_n(\lambda), \beta_1(\lambda), \beta_2(\lambda), \cdots, \beta_n(\lambda)$, such that
$$\alpha(\lambda) = (\alpha_1(\lambda), \alpha_2(\lambda), \cdots, \alpha_n(\lambda))y_0, \tag{1.14.7}$$
$$\beta(\lambda) = (\beta_1(\lambda), \beta_2(\lambda), \cdots, \beta_n(\lambda))y_0. \tag{1.14.8}$$

The approach to evaluate $\alpha_1(0), \alpha_2(0), \cdots, \alpha_n(0), \beta_1(0), \beta_2(0), \cdots, \beta_n(0)$ has been given in §1.9, where it is aimed at the plane elasticity problems. For $\lambda \neq 0$ we may use the following power series approach.

We make a similarity transformation $x \to \xi^{-1}x, y \to \xi^{-1}y$, then Γ_1 is transferred to Γ_0, and the equation (1.14.1) becomes
$$-\triangle u + \lambda\xi^2 u = 0. \tag{1.14.9}$$

Let $z_k = y_{k+1}(k = 0, 1, \cdots)$, then z_k is the infinite element solution to the equation (1.14.9). By (1.14.6) we get
$$z_k \sim \alpha(\lambda)g_1 + \lambda_2\beta(\lambda)g_2\lambda_2^k.$$

Then (1.14.7),(1.14.8) yield
$$\alpha(\lambda) = (\alpha_1(\lambda\xi^2), \alpha_2(\lambda\xi^2), \cdots, \alpha_n(\lambda\xi^2))z_0, \tag{1.14.10}$$
$$\lambda_2\beta(\lambda) = (\beta_1(\lambda\xi^2), \beta_2(\lambda\xi^2), \cdots, \beta_n(\lambda\xi^2))z_0. \tag{1.14.11}$$

(1.14.7) and (1.14.10) lead to
$$(\alpha_1(\lambda), \alpha_2(\lambda), \cdots, \alpha_n(\lambda)) = (\alpha_1(\lambda\xi^2), \alpha_2(\lambda\xi^2), \cdots, \alpha_n(\lambda\xi^2))X(\lambda), \tag{1.14.12}$$
and (1.14.8) and (1.14.11) lead to
$$\lambda_2(\beta_1(\lambda), \beta_2(\lambda), \cdots, \beta_n(\lambda)) = (\beta_1(\lambda\xi^2), \beta_2(\lambda\xi^2), \cdots, \beta_n(\lambda\xi^2))X(\lambda).$$

1.14 HELMHOLTZ EQUATION

We set
$$(\alpha_1(\lambda), \alpha_2(\lambda), \cdots, \alpha_n(\lambda)) = \sum_{m=0}^{\infty} a_m \lambda^m$$

in the neighborhood of $\lambda = 0$, where a_m are n-dimensional row vectors. Substituting it into (1.14.12) and noting (1.14.5) we obtain

$$\sum_{m=0}^{\infty} a_m \lambda^m = \sum_{m=0}^{\infty} a_m (\lambda \xi^2)^m \sum_{m=0}^{\infty} \lambda^m X_m = \sum_{m=0}^{\infty} \lambda^m \sum_{i=0}^{m} \xi^{2i} a_i X_{m-i}.$$

We get a system of equations

$$a_m = \sum_{i=0}^{m} \xi^{2i} a_i X_{m-i}, \quad m = 1, 2, \cdots,$$

that is

$$a_m = \left(\sum_{i=0}^{m-1} \xi^{2i} a_i X_{m-i} \right) (I - \xi^{2m} X_0)^{-1}, \quad m = 1, 2, \cdots.$$

In the same way, let

$$(\beta_1(\lambda), \beta_2(\lambda), \cdots, \beta_n(\lambda)) = \sum_{m=0}^{\infty} b_m \lambda^m,$$

then

$$b_m = \left(\sum_{i=0}^{m} \xi^{2i} b_i X_{m-i} \right) (\lambda_2 I - \xi^{2m} X_0)^{-1}, \quad m = 1, 2, \cdots.$$

We turn now to the eigenvalue problems associated with (1.14.1). Let O be a corner point on the boundary of the domain Ω_0. Ω_0 is decomposed into Ω^* and Ω, where Ω includes the point O. Ω is divided into infinite similar triangular elements and Ω^* is divided into finite conventional triangular elements (Fig.22). For simplicity we assume that the boundary codition $u = 0$ is given on the whole boundary. Let the eigenvalue with minimum absolute value be μ, and the corresponding eigenfunction be u_1, so they satisfy on Ω_0 the equation

$$-\Delta u_1 + \mu u_1 = 0. \tag{1.14.13}$$

Let Ω_ξ be a domain surrounded by Γ_1, Γ^* and Γ_*. We consider an analogous eigenvalue problem on Ω or Ω_ξ instead of the domain Ω_0, and the corresponding eigenvalues are denoted by μ_1 and μ_2. We will prove in the next chapter that μ, μ_1, μ_2 are all negative numbers and

$$|\mu| \leq |\mu_1| < |\mu_2|. \tag{1.14.14}$$

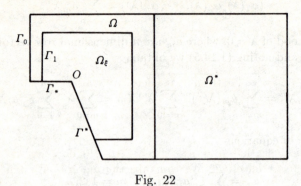

Fig. 22

First of all we assume that $|\mu| < |\mu_1|$, then we can restrict λ such that $|\lambda| < |\mu_1|$, thus

$$y_1 = X(\lambda)y_0. \qquad (1.14.15)$$

Let the nodal values on $\bar{\Omega}^* \setminus \partial\Omega_0$ be arranged into a vector y^* and the stiffness matrix on Ω^* be $K^*(\lambda)$, then the strain energy on the union of Ω^* and the first layer is

$$W(\lambda) = \frac{1}{2}(y^*)^T K^*(\lambda) y^* + \frac{1}{2}\begin{pmatrix} y_0^T & y_1^T \end{pmatrix} \begin{pmatrix} K_0 + \lambda L_0 & -A^T - \lambda D^T \\ -A - \lambda D & K_0' + \lambda L_0' \end{pmatrix} \begin{pmatrix} y_0 \\ y_1 \end{pmatrix}, \qquad (1.14.16)$$

where part of the components of y_0 vanish, and the rest part of components are included in y^*.

Now we assume that u_1 is the infinite element solution to (1.14.13), then we have $\frac{\partial W(\lambda)}{\partial y^*} = 0$. (1.14.16) is a quadratic form including y^* and y_1, hence its derivative is a linear form

$$\frac{\partial W(\lambda)}{\partial y^*} = A(\lambda)y^* + B(\lambda)y_1,$$

where $A(\lambda)$ and $B(\lambda)$ are matrices depending on λ. Substituting (1.14.15) into it, we get

$$A(\lambda)y^* + B(\lambda)X(\lambda)y_0 = 0,$$

or

$$C(\lambda)y^* = 0, \qquad (1.14.17)$$

where $C(\lambda)$ is a square matrix depending on λ. y^* is the nontrivial solution to the equation (1.14.17). Therefore μ is the root of equation $\det C(\mu) = 0$. The roots of it are not unique, we take the one with minimum absolute value, and let it be μ^*. Since μ is a root, $|\mu^*| \leq |\mu|$. On the other hand we can derive $\frac{\partial W(\mu^*)}{\partial y^*} = 0$ from (1.14.17), so μ^* is an eigenvalue, hence $\mu^* = \mu$.

1.14 HELMHOLTZ EQUATION

To evaluate μ, the quasi-Newtonian iterative scheme can be applied. Taking
$$\mu^{(0)} = 1, \mu^{(1)} = 0,$$
we use the following iterative scheme,
$$\mu^{(m+1)} = \mu^{(m)} - \frac{\det C(\mu^{(m)})}{\det C(\mu^{(m)}) - \det C(\mu^{(m-1)})}(\mu^{(m)} - \mu^{(m-1)}),$$
$$m = 1, 2, \cdots,$$
then μ is obtained.

Secondly we assume that $\mu = \mu_1$. Then (1.14.15) does not hold. The above approach ceases to be effective. We can take the union of Ω^* and the first layer to be a new domain Ω^*, and the second layer to be the new first layer. There is a particular case:

If $\Omega_0 = \Omega$, then the strain energy on the union of the first and the second layers is
$$W(\lambda) = \frac{1}{2} \begin{pmatrix} y_1^T & y_2^T \end{pmatrix} \begin{pmatrix} K_0 + \lambda \xi^2 L_0 & -A^T - \lambda \xi^2 D^T \\ -A - \lambda \xi^2 D & K_0' + \lambda \xi L_0' \end{pmatrix} \begin{pmatrix} y_1 \\ y_2 \end{pmatrix}$$
$$+ \frac{1}{2} y_1^T (K_0' + \lambda L_0') y_1.$$

Letting $\frac{\partial W(\lambda)}{\partial y_1} = 0$ we obtain
$$(K + \lambda L) y_1 - (A^T + \lambda \xi^2 D^T) y_2 = 0.$$

Substituting
$$y_2 = X(\lambda \xi^2) y_1$$
into it, we get
$$\{(K + \lambda L) - (A^T + \lambda \xi^2 D^T) X(\lambda \xi^2)\} y_1 = 0.$$

The matrix of the coefficients is the same as that of the equation (1.14.4). We conclude that λ is just the eigenvalue as the matrix of the coefficients of the equation (1.14.4) is singular.

The Helmholtz equation is an example for those equations where the stiffness matrices are different on different layers. We overcome this difficulty by using the properties of analytic functions. However the approach in this section is not sufficient for general elliptic equations. We will consider more general cases in the next section. Next section is also a generalization of the content in §1.11, where a nonhomogeneous term in the equations causes that the important relation (1.1.7) can not be used directly.

1.15 Elliptic equations with variable coefficients

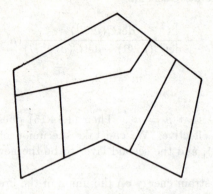

Fig. 23

Let Ω_0 be a polygonal domain. There are finite many broken lines which divide it into finite polygonal subdomains $\Omega_l, l = 1, \ldots, L$ (Fig.23). We assume that $a_{ij}(x,y)$, $1 \leq i,j \leq 2$, are differentiable functions, $a_{12}(x,y) = a_{21}(x,y)$, and the elliptic condition

$$a_{11}\eta_1^2 + 2a_{12}\eta_1\eta_2 + a_{22}\eta_2^2 \geq \alpha(\eta_1^2 + \eta_2^2)$$

is satisfied for all real numbers η_1, η_2, where the constant $\alpha > 0$. Besides, we assume that a function $p(x,y)$ takes a positive constant value on each subdomain Ω_l and a function $a_0(x,y)$ is non-negative. We intend to solve the following Dirichlet problem,

$$-\frac{\partial}{\partial x}\left(a_{11}p\frac{\partial u}{\partial x} + a_{12}p\frac{\partial u}{\partial y}\right) - \frac{\partial}{\partial y}\left(a_{21}p\frac{\partial u}{\partial x} + a_{22}p\frac{\partial u}{\partial y}\right)$$
$$+ a_0 u = f(x,y), \qquad (1.15.1)$$

$$u\big|_{\partial\Omega_0} = 0, \qquad (1.15.2)$$

where $\partial\Omega_0$ is the boundary of Ω_0. The Poisson equation and the Helmholtz equation considered in the preceding sections are special cases of (1.15.1).

Since the coefficients are discontinuous, u satisfies (1.15.1) in a general sense. It satisfies the equation classically in each subdomain, and on interfaces it satisfies the following conditions:

$$\begin{cases} u \text{ is continuous on interfaces,} \\ p\left(a_{11}\nu_x\frac{\partial u}{\partial x} + a_{12}\nu_x\frac{\partial u}{\partial y} + a_{21}\nu_y\frac{\partial u}{\partial x} + a_{22}\nu_y\frac{\partial u}{\partial y}\right) \text{ is continuous on interfaces,} \end{cases}$$

where ν_x, ν_y are the components of the unit normal vectors on interfaces. The solution u may possess singularities at the following points: the cross points of interfaces, the turning points of interfaces, the cross points of interfaces with the boundary $\partial\Omega_0$, and the points on $\partial\Omega_0$ with reentrant interior angles. The first-order derivatives of the solution near these points, which will be generally known as singular points, may be unbounded.

We will prove in the next chapter that near each singular point b_1 the solution can be decomposed as $u = v + w$, where v satisfies the equation

$$-\frac{\partial}{\partial x}\left(a_{11}(b_1)p\frac{\partial u}{\partial x} + a_{12}(b_1)p\frac{\partial u}{\partial y}\right) - \frac{\partial}{\partial y}\left(a_{21}(b_1)p\frac{\partial u}{\partial x} + a_{22}(b_1)p\frac{\partial u}{\partial y}\right) = 0, \tag{1.15.3}$$

and w is a "good" function without singularity. The coefficients of the equation (1.15.3) are piecewise constants, so our previous approaches can be applied to this equation. The above decomposition suggests us using two sets of elements, finite and infinite, on the same domain Ω. We take triangular elements as usual, and the infinite element partition is described in §1.7. We require that two sets of elements conform to each other, that is, the nodes of them coincide on the boundary of Ω (Fig.24).

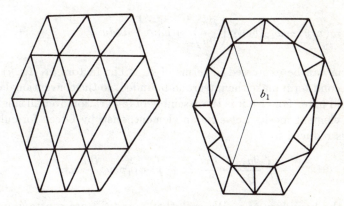

Fig. 24

We turn now to derive the combined stiffness matrix on Ω. If v is the infinite element solution to the equation (1.15.3), then we can define vectors $y_k, k = 0, 1, \ldots,$ of nodal values and obtain the transfer matrix X and the combined stiffness matrix K_z, and if w is a piecewise linear function defined on the finite element mesh, then the strain energy is

$$W = \frac{1}{2}\int_\Omega \left\{ a_{11}p\left(\frac{\partial(v+w)}{\partial x}\right)^2 + 2a_{12}p\frac{\partial(v+w)}{\partial x}\frac{\partial(v+w)}{\partial y} \right.$$
$$\left. + a_{22}p\left(\frac{\partial(v+w)}{\partial y}\right)^2 + a_0(v+w)(v+w) \right\} dxdy.$$

To evaluate the stiffness matrix, three kinds of integrations should be taken into account:

(a) $\int_\Omega \left\{ a_{11} p \left(\frac{\partial v}{\partial x} \right)^2 + 2 a_{12} p \frac{\partial v}{\partial x} \frac{\partial v}{\partial y} + a_{22} p \left(\frac{\partial v}{\partial y} \right)^2 + a_0 v^2 \right\} dxdy,$

(b) $2 \int_\Omega \left\{ a_{11} p \frac{\partial v}{\partial x} \frac{\partial w}{\partial x} + a_{12} p \frac{\partial v}{\partial x} \frac{\partial w}{\partial y} + a_{12} p \frac{\partial w}{\partial x} \frac{\partial v}{\partial y} + a_{22} p \frac{\partial v}{\partial y} \frac{\partial w}{\partial y} + a_0 vw \right\} dxdy,$

(c) $\int_\Omega \left\{ a_{11} p \left(\frac{\partial w}{\partial x} \right)^2 + 2 a_{12} p \frac{\partial w}{\partial x} \frac{\partial w}{\partial y} + a_{22} p \left(\frac{\partial w}{\partial y} \right)^2 + a_0 w^2 \right\} dxdy.$

For (a), it can be written as

$$\int_\Omega \left\{ a_{11}(b_1) p \left(\frac{\partial v}{\partial x} \right)^2 + 2 a_{12}(b_1) p \frac{\partial v}{\partial x} \frac{\partial v}{\partial y} + a_{22}(b_1) p \left(\frac{\partial v}{\partial y} \right)^2 \right\} dxdy$$

$$+ \int_\Omega \left\{ (a_{11}(x,y) - a_{11}(b_1)) p \left(\frac{\partial v}{\partial x} \right)^2 + 2(a_{12}(x,y) - a_{12}(b_1)) p \frac{\partial v}{\partial x} \frac{\partial v}{\partial y} \right.$$

$$\left. + (a_{22}(x,y) - a_{22}(b_1)) p \left(\frac{\partial v}{\partial y} \right)^2 + a_0 v^2 \right\} dxdy.$$

The first term can be expressed in terms of K_z. The factors $(a_{11}(x,y) - a_{11}(b_1))$ etc. in the second term make the integrand bounded, so the conventional quadrature scheme is valid here. For (b), it is the assembly of integration over all single elements of the finite element mesh. Let e be an element, then by Green formula, the first term of (b) is reduced to

$$- \int_e \frac{\partial a_{11}}{\partial x} pv \frac{\partial w}{\partial x} dx + \int_{\partial e} a_{11} p v \nu_x \frac{\partial w}{\partial x} ds,$$

where ∂e is the boundary of e. We note that p and $\frac{\partial w}{\partial x}$ are constants on e, so the integrands of the two integrations are continuous and bounded, and the conventional quadrature scheme is valid here. The treatment of the other terms of (b) is just the same. (c) is just the conventional finite element stiffness matrices evaluation.

Let z_1 be a vector consisting of boundary nodal values of the finite element mesh, and z_2 be a vector consisting of interior nodal values. We can write down the strain energy as

$$W = \frac{1}{2} (y_0^T \ z_1^T \ z_2^T) \begin{pmatrix} L_{11} & L_{12} & L_{13} \\ L_{12}^T & L_{22} & L_{23} \\ L_{13}^T & L_{23}^T & L_{33} \end{pmatrix} \begin{pmatrix} y_0 \\ z_1 \\ z_2 \end{pmatrix}, \qquad (1.15.4)$$

where y_0 and z_1 correspond to the same nodes. If the boundary data are given then

$$z_1 = g - y_0, \qquad (1.15.5)$$

where g is the vector of boundary data. We require that W reaches its minimum for fixed g, so we take first order derivatives of (1.15.4) with respect to y_0 restricted by (1.15.5) and obtain

$$\frac{\partial W}{\partial y_0} = (L_{11} - L_{12}^T)y_0 + (L_{12} - L_{22})z_1 + (L_{13} - L_{23})z_2 = 0.$$

Noting (1.15.5) we have

$$(L_{11} - L_{12}^T - L_{12} + L_{22})y_0 + (L_{12} - L_{22})g + (L_{13} - L_{23})z_2 = 0, \qquad (1.15.6)$$

which leads to

$$y_o = (L_{11} - L_{12}^T - L_{12} + L_{22})^{-1}((L_{22} - L_{12})g + (L_{23} - L_{13})z_2). \qquad (1.15.7)$$

We substitute (1.15.5) and (1.15.7) into (1.15.4). After a few calculation we get

$$W = \frac{1}{2}\begin{pmatrix}g^T & z_2^T\end{pmatrix}\begin{pmatrix}K_{z1} & K_{z2} \\ K_{z2}^T & K_{z3}\end{pmatrix}\begin{pmatrix}g \\ z_2\end{pmatrix}, \qquad (1.15.8)$$

where

$$K_{z1} = L_{22} - (L_{22} - L_{12}^T)(L_{11} - L_{12}^T - L_{12} + L_{22})^{-1}(L_{22} - L_{12}),$$

$$K_{z2} = L_{23} - (L_{22} - L_{12}^T)(L_{11} - L_{12}^T - L_{12} + L_{22})^{-1}(L_{23} - L_{13}),$$

$$K_{z3} = L_{33} - (L_{23} - L_{13}^T)(L_{11} - L_{12}^T - L_{12} + L_{22})^{-1}(L_{23} - L_{13}).$$

We assemble the combined stiffness matrices given in (1.15.8) for all singular points with the stiffness matrices of the other elements, then we can solve the problem on the entire domain, or we may use the matrix in (1.15.4) with two constraints (1.15.5) and (1.15.6).

1.16 Parabolic equations

Let us consider the same domain and partition as the previous section. This section is devoted to the initial-boundary value problems of the parabolic equation,

$$\frac{\partial u}{\partial t} - \frac{\partial}{\partial x}\left(a_{11}p\frac{\partial u}{\partial x} + a_{12}p\frac{\partial u}{\partial y}\right) - \frac{\partial}{\partial y}\left(a_{21}p\frac{\partial u}{\partial x} + a_{22}p\frac{\partial u}{\partial y}\right) + a_0 u = f,$$

$$u\big|_{\partial\Omega_0} = 0,$$

$$u\big|_{t=0} = u_0.$$

We assume that the implicit Euler's scheme is applied. Let Δt be the length of time step, and $u^{(m)} = u(x, y, m\Delta t)$, then the scheme reads

$$u^{(0)} = u_0,$$

$$\frac{u^{(m)} - u^{(m-1)}}{\Delta t} - \frac{\partial}{\partial x}\left(a_{11}p\frac{\partial u^{(m)}}{\partial x} + a_{12}p\frac{\partial u^{(m)}}{\partial y}\right)$$
$$- \frac{\partial}{\partial y}\left(a_{21}p\frac{\partial u^{(m)}}{\partial x} + a_{22}p\frac{\partial u^{(m)}}{\partial y}\right) + a_0 u^{(m)} = f^{(m)}, \quad m = 1, 2, \cdots,$$

$$u^{(m)}\big|_{\partial\Omega_0} = 0,$$

where $f^{(m)} = f(x, y, m\Delta t)$. The equations can be rewritten as

$$-\frac{\partial}{\partial x}\left(a_{11}p\frac{\partial u^{(m)}}{\partial x} + a_{12}p\frac{\partial u^{(m)}}{\partial y}\right) - \frac{\partial}{\partial y}\left(a_{21}p\frac{\partial u^{(m)}}{\partial x} + a_{22}p\frac{\partial u^{(m)}}{\partial y}\right)$$
$$+ \left(a_0 + \frac{1}{\Delta t}\right)u^{(m)} = f^{(m)} + \frac{1}{\Delta t}u^{(m-1)}.$$

Therefore the approach of the previous section can be applied here with a_0 replaced by $(a_0 + \frac{1}{\Delta t})$ and f by $(f^{(m)} + \frac{1}{\Delta t}u^{(m-1)})$.

1.17 Variational inequalities

We consider the same domain and partition as those in §1.13. But we make one more finite element partition in the domain Ω since we will solve a problem where the solution u is more complicated than a harmonic function in Ω. Associated with the Laplace operator the total energy is given by

$$J(u) = \frac{1}{2}\int_{\Omega_0}\left\{\left(\frac{\partial u}{\partial x}\right)^2 + \left(\frac{\partial u}{\partial y}\right)^2\right\}dxdy - \int_{\Omega_0} fu\,dxdy. \tag{1.17.1}$$

We assume that u satisfies the boundary condition,

$$u\big|_{\partial\Omega_0} = 0. \tag{1.17.2}$$

And we define an "obstacle" function $\varphi(x, y)$, which is continuous on Ω_0 and assumes non-positive values on the boundary $\partial\Omega_0$. We require that $u \geq \varphi$, u satisfies (1.17.2), and

$$J(u) = \min_{v \geq \varphi, v|_{\partial\Omega_0} = 0} J(v).$$

We solve the above problem by means of the infinite element method. Let z_3 be a vector consisting of all nodal values in the domain Ω^*, then using the approach and notations in §1.15 we have

$$J(u) = \frac{1}{2} \begin{pmatrix} g^T & z_2^T \end{pmatrix} \begin{pmatrix} K_{z1} & K_{z2} \\ K_{z2}^T & K_{z3} \end{pmatrix} \begin{pmatrix} g \\ z_2 \end{pmatrix} + \frac{1}{2} \begin{pmatrix} g^T & z_3^T \end{pmatrix} K \begin{pmatrix} g \\ z_3 \end{pmatrix}$$

$$+ \begin{pmatrix} g^T & z_2^T & z_3^T \end{pmatrix} L\tilde{f}, \tag{1.17.3}$$

where K is the total stiffness matrix on Ω^*, the third term comes from the second term of (1.17.1), and the vector $\tilde{f}$ is obtained from the nodal values of f. It should be noticed that the function fu is integrated twice on the domain Ω. For the infinite element mesh, y_0 is given by (1.15.7), and for the finite element mesh, z_1 is given by (1.15.5).

Now let us consider the constraints. Let e be a single element of the finite element mesh in Ω. Let the mean value of φ on e be φ_e. We set $u = v + w$, where v is the interpolation function of the infinite element mesh and w is the interpolation function of the finite element mesh. We evaluate the mean values v_e, w_e on e, where v_e depends on y_0 and w_e depends on z_2 and z_1. We set $u_e = v_e + w_e$. If $e \subset \Omega^*$, it is easier to obtain the mean value u_e. We make the constraints as

$$u_e \geq \varphi_e \tag{1.17.4}$$

on all elements. Subjected to (1.17.4), the minimum value problem of the function (1.17.3) is the infinite element approximation to the original problem.

Notes to Chapter 1

The content of this chapter is based on [4],[7],[8],[10],[17],[19],[20],[25],[27],[30], [31],[33],[34],[35],[38],[39] and [41] with significant reorganization, some new material is added, and only algorithm is discussed in this chapter, the related proof is postponed until the next chapter.

The content of §1.1–§1.3 represents three typical methods, the eigenvalue method, the Fourier method and the iterative method. They were developed in [4],[17],[7],[31] and [19]. For the sake of reading convenience, we only discuss one typical example in those three sections, which is the two dimensional exterior problem of the Laplace equation. The infinite element method for this problem first appeared in [5] and [10]. To be a beginning this problem is suitable, because the equation is simple, the algorithm has nothing to do with the boundary condition, and the Fourier method is especially applicable for this problem.

We consider the infinite element method for various equations and various boundary conditions in §1.4–§1.8 and §1.10. Since the same three methods are applied to these different problems, we emphasize the specific character of each individual in the process of solving it in these sections. We hope that these sections enable the readers not only to know the infinite element method for a few boundary value problems, but also to draw inferences about other cases. For other equations and other boundary conditions which we have not touched upon, the readers can also handle them with reference to the methods in these sections. The content of these sections is partly of the following origin, the isoparametric elements first appeared in [8], the corner problems in [2] and [4], the plane elasticity problems in [4], and the exterior Stokes problems in [26].

In §1.9 and §1.13 there is some material closely related to the above treatment for different problems. §1.9 is an application of §1.8, which is based on [4] and [31]. §1.13 is about how to solve the algebraic system generated from the infinite element method, where we study the multigrid method which is popular in the recent years. The material of this section is based on [34] and [67].

In §1.11, §1.12, §1.14, and §1.15 we present some new approaches which enable us to deal with a great variety of problems with the infinite element method. The material of §1.11 and §1.12 is based on [10] and [35], §1.14 on [20] and [30], and §1.15 on [33]. §1.16 and §1.17 are some extensions of §1.15.

2   Foundations of Algorithm

2.1 Infinite element spaces

We are going to prove in this chapter the formulas and conclusions appeared in Chapter One. In what follows we will denote by C a generic constant which may stand for different values at different places. If we need to distinguish some different constants, then they are denoted by C_1, C_2, etc. Let $x = (x_1, \ldots, x_d)$ be a point in the space $\mathbb{R}^d, d = 2, 3$, and $\Omega \subset \mathbb{R}^d$ be an open set with boundary $\partial \Omega$. Let $C_0^\infty(\Omega)$ be the set of functions with compact support which are infinitely differentiable on Ω. The normal notations $H^s(\Omega)$ and $W^{m,p}(\Omega)$ for the Sobolev Spaces along with the notations of norms and seminorms,

$$|u|_{k,p,\Omega} = \left\{ \int_\Omega |\partial^k u|^p \, dx \right\}^{\frac{1}{p}}, \quad k \geq 0, p \geq 1,$$

$$|u|_{k,\infty,\Omega} = \operatorname*{esssup}_{x \in \Omega} |\partial^k u|, \quad k \geq 0,$$

$$\|u\|_{m,p,\Omega} = \left\{ \sum_{k=0}^m |u|_{k,p,\Omega}^p \right\}^{\frac{1}{p}}, \quad m \geq 0, p \geq 1,$$

$$\|u\|_{m,\infty,\Omega} = \max_{0 \leq k \leq m} |u|_{k,\infty,\Omega},$$

are applied, where

$$|\partial^k u| = \sum_{|\alpha|=k} \left| \frac{\partial^k u}{\partial x_1^{\alpha_1} \cdots \partial x_d^{\alpha_d}} \right|, \quad |\alpha| = \alpha_1 + \cdots + \alpha_d.$$

If $p = 2$, the above norms and seminorms are denoted by $\|u\|_{m,\Omega}$ and $|u|_{k,\Omega}$ respectively. For any real number s, the space $H^s(\Omega)$ is equipped with the norm $\|\cdot\|_{s,\Omega}$. We will omit the symbol Ω in those notations when there is no confusion. Finally the L^2-inner product is denoted by $(\cdot, \cdot)$.

For an unbounded domain $\Omega \subset \mathbb{R}^3$, if $u \in C_0^\infty(\Omega)$, then we have the inequality [53],

$$\int_\Omega \frac{u^2(x)}{|x-y|^2} \, dx \leq 4 \int_\Omega |\nabla u(x)|^2 \, dx, \tag{2.1.1}$$

where $y \in \mathbb{R}^3$ is an arbitrary point and $|\cdot|$ stands for the Euclidean norm of a vector. For a plane domain $\Omega = \{1 < |x| < \infty\} \subset \mathbb{R}^2$, if $u \in C_0^\infty(\Omega)$, then we have the inequality [53],

$$\int_{|x|>1} \frac{u^2(x)}{|x|^2 \log^2 |x|} \, dx \le 4 \int_{|x|>1} |\nabla u(x)|^2 \, dx. \tag{2.1.2}$$

Inequalities (2.1.1) and (2.1.2) are very useful in considering various bounday value problems on unbounded domains.

For an exterior domain $\Omega \subset \mathbb{R}^2$, without losing generality we always assume that it is included in $\{1 < |x| < \infty\}$. We assume that the boundary $\partial \Omega$ is a simply closed curve in $\mathbb{R}^2$ and satisfies the Lipschitz condition. Denote by $\bar{\Omega}$ the closure of Ω. The set $C_0^\infty(\bar{\Omega})$ consists of all infinitely differentiable functions on $\bar{\Omega}$ which have compact support, and the norm for every $u \in C_0^\infty(\bar{\Omega})$ is defined as

$$\|u\|_{1,*} = \left(\int_\Omega \left(|\nabla u(x)|^2 + \frac{u^2(x)}{|x|^2 \log^2 |x|} \right) dx \right)^{\frac{1}{2}},$$

then $C_0^\infty(\bar{\Omega})$ admits an associated inner product. The completion of it is a Hilbert space, denoted by $H^{1,*}(\Omega)$.

Let a three dimensional exterior domain be $\Omega \subset \mathbb{R}^3$. Without losing generality we always assume that $O \notin \bar{\Omega}$. Space $H^{1,*}(\Omega)$ is equipped with the norm

$$\|u\|_{1,*} = \left(\int_\Omega \left(|\nabla u(x)|^2 + \frac{u^2(x)}{|x|^2} \right) dx \right)^{\frac{1}{2}}.$$

The trace of $u \in H^{1,*}(\Omega)$ on $\partial \Omega$ is denoted by $u\big|_{\partial \Omega}$. We define a subspace of $H^{1,*}(\Omega)$,

$$H_0^{1,*}(\Omega) = \{u \in H^{1,*}(\Omega); u\big|_{\partial \Omega} = 0\}.$$

The following lemma gives an outline of the space $H^{1,*}(\Omega)$ for dimensions $d = 2, 3$.

Lemma 2.1.1. *The equivalent definition for $H^{1,*}(\Omega)$ is*

$$H^{1,*}(\Omega) = \{u \in H_{\text{loc}}^1(\Omega); \|u\|_{1,*} < +\infty\}. \tag{2.1.3}$$

Moreover if $d = 2$, then

$$H^{1,*}(\Omega) = \{u \in H_{\text{loc}}^1(\Omega); \|\nabla u\|_0 < +\infty\} \tag{2.1.4}$$

also holds.

Proof. Let the right hand side of (2.1.3) be D. Evidently $H^{1,*}(\Omega) \subset D$. We prove $D \subset H^{1,*}(\Omega)$. Let χ be a function which belongs to $C^\infty(\Omega)$, and it is equal to 1 for large $|x|$, and vanishes near $\partial \Omega$. If $u \in D$, we set $u_1 = \chi u$. It is easy to see $(1-\chi)u \in H^{1,*}(\Omega)$, so it will suffice to varify $u_1 \in H^{1,*}(\Omega)$. We take a constant b, $b \ge 2$, then

$$\int_b^{+\infty} \frac{d\xi}{\xi \log \xi} = +\infty.$$

So we can take another constant c, such that

$$\int_b^c \frac{d\xi}{\xi \log \xi} = 1.$$

Difine a function

$$f_b(x) = \begin{cases} 1, & |x| \le b, \\ 1 - \int_b^{|x|} \frac{d\xi}{\xi \log \xi}, & b < |x| < c, \\ 0, & |x| \ge c. \end{cases}$$

f_b satisfies

$$|\nabla f_b(x)| \le \frac{1}{|x| \log |x|}.$$

Let $v = f_b u_1$, then

$$\nabla v = u_1 \nabla f_b + f_b \nabla u_1.$$

Hence

$$\int_{|x|>b} |\nabla v|^2 \, dx \le \int_{|x|>b} \left(|\nabla u_1(x)|^2 + \frac{u_1^2(x)}{|x|^2 \log^2 |x|} \right) dx \to 0 \quad (b \to \infty).$$

Consequently

$$\int_\Omega |\nabla(u_1 - v)|^2 \, dx = \int_{|x|>b} |\nabla(u_1 - v)|^2 \, dx \to 0 \quad (b \to \infty).$$

For fixed b the function v belongs to $H^1(\Omega)$ and the support of v is compact in Ω. Thus there is a function $w \in C_0^\infty(\Omega)$ such that

$$\int_\Omega |\nabla(v - w)|^2 \, dx < \varepsilon$$

for any $\varepsilon > 0$. Consequently

$$\int_\Omega |\nabla(u_1 - w)|^2 \, dx < \varepsilon$$

for any $\varepsilon > 0$. We choose a sequence $\{w_k\} \subset C_0^\infty(\Omega)$ such that

$$\int_\Omega |\nabla(u_1 - w_k)|^2 \, dx \to 0 \quad (k \to \infty).$$

Then (2.1.1) or (2.1.2) implies

$$\|w_k - w_l\|_{1,*} \to 0 \quad (k \to \infty, l \to \infty),$$

which means $\{w_k\}$ is a Cauchy sequence in $H^{1,*}(\Omega)$. Therefore $u_1 \in H^{1,*}(\Omega)$, which proves (2.1.3).

We turn now to the case of $d = 2$. If $\|\nabla u\|_0 < +\infty$, we set $u_1 = \chi u$. It will suffice to prove $u_1 \in H^{1,*}(\Omega)$. Suppose the support of u_1 is included in $|x| > b$, we make a reflection transform and let $w(x) = u_1\left(\frac{x}{|x|^2}\right)$, then w is defined on $\Omega' = \{x; 0 < |x| < \frac{1}{b}\}$. The set $\{0\}$ is a (2.1.2)-polar set [60], hence $C_0^\infty(\Omega')$ is dense in $\{u \in H^1(\Omega'); u|_{|x|=\frac{1}{b}} = 0\}$. For any $\varepsilon > 0$ there is a $w_1 \in C_0^\infty(\Omega')$ such that $\|w_1 - w\|_{1,\Omega'} < \varepsilon$. Let $v(x) = w_1\left(\frac{x}{|x|^2}\right)$, and $\Omega'' = \{x; b < |x| < \infty\}$, then $\|\nabla(u_1 - v)\|_{0,\Omega''} < \varepsilon$. Following the argument to prove (2.1.3), we obtain (2.1.4). □

Now let $\Omega \subset \mathbb{R}^2$, and we consider the infinite element subspaces of $H^{1,*}(\Omega)$. $\partial\Omega = \Gamma_0$ is assumed to be star-shape with respect to the point O, that is the line segment linking each point on Γ_0 with the point O is included entirely in the interior of Γ_0. We divide the domain Ω into triangular elements according to the approach stated in §1.1, then denote by $\xi^k\Omega$ the exterior domain of Γ_k, and let $\Omega_k = \xi^{k-1}\Omega \setminus \overline{\xi^k\Omega}$. Let $e_1, e_2, \ldots$ be the triangular elements. Denote by $P_m(\Omega')$ the set of all polynomials in x of degree $\leq m$, where Ω' is an arbitrary open set. We define the infinite element spaces as follows:

$$S(\Omega) = \left\{u \in H^{1,*}(\Omega); u|_{e_i} \in P_1(e_i), i = 1, 2, \ldots\right\},$$

$$S_0(\Omega) = \left\{u \in S(\Omega); u|_{\partial\Omega} = 0\right\},$$

$$S(\Gamma_k) = \left\{u|_{\Gamma_k}; u \in S(\Omega)\right\}.$$

For $u \in S(\Omega)$ we arrange the nodal values on Γ_k into a column vector y_k as we did in Chapter One. Let $y_k = B_k u$. Clearly each element in $S(\Gamma_k)$ corresponds to one $y_k \in \mathbb{R}^n$. We will identify y_k with $u|_{\Gamma_k}$ rather than distinguish them.

Lemma 2.1.2. If $u \in S(\Omega)$, $y_k = B_k u$, then there is a constant C independent of u, such that

$$\frac{1}{C}\sum_{k=1}^\infty |y_k - y_{k-1}|^2 \leq \int_\Omega |\nabla u|^2 \, dx \leq C\sum_{k=0}^\infty |y_k|^2. \tag{2.1.5}$$

Proof. Let $e_j \subset \Omega_k$ be an arbitrary triangular element. It is shown in Fig. 2 that one side of e_j points to the point O. Let the length of this side be L. Let meas e_j denote the area of the triangle e_j, and ρ the least height of the triangle e_j. Let the values of u at the both ends of the side pointing to the point O be $y_{k-1}^{(i)}$ and $y_k^{(i)}$, then

$$\frac{|y_k^{(i)} - y_{k-1}^{(i)}|}{L} \leq |\nabla u| \leq \frac{|y_k| + |y_{k-1}|}{\rho}.$$

Therefore
$$\frac{\operatorname{meas} e_j}{L^2}|y_k^{(i)} - y_{k-1}^{(i)}|^2 \leq \int_{e_j} |\nabla u|^2 \, dx \leq \frac{\operatorname{meas} e_j}{\rho^2}(|y_k| + |y_{k-1}|)^2.$$

By similarity $\frac{\operatorname{meas} e_j}{L^2}$ and $\frac{\operatorname{meas} e_j}{\rho^2}$ depend only on the position of e_j in Ω_k and are independent of k. Summing up the above inequalities with respect to all triangles e_j we obtain (2.1.5). □

Corresponding to the Laplace operator we define the following bilinear functional on $S(\Omega)$,
$$a(u,v) = \int_{\Omega} \nabla u \cdot \nabla v \, dx, \tag{2.1.6}$$
where $\nabla u \cdot \nabla v$ is the inner product of ∇u and ∇v. Sometimes we write $a(u,v)_\Omega$ for this bilinear functional to indicate the domain of integration.

Now we have the discrete variational problem: find $u \in S(\Omega)$ for any $f \in S(\partial\Omega)$ such that $u|_{\partial\Omega} = f$ and
$$a(u,v) = 0, \qquad \forall v \in S_0(\Omega). \tag{2.1.7}$$

Lemma 2.1.3. *Problem (2.1.7) admits a unique solution.*

Proof. We take a $u_0 \in S(\Omega)$ such that $u_0|_{\partial\Omega} = f$ and $u_0|_{\Gamma_k} = 0$, $k = 1, 2, \cdots$. Let $u' = u - u_0$, then the problem (2.1.7) is equivalent to: find $u' \in S_0(\Omega)$ such that
$$a(u', v) = -a(u_0, v), \qquad \forall v \in S_0(\Omega). \tag{2.1.8}$$
The inequality (2.1.2) holds on $S_0(\Omega)$, hence
$$a(v,v) \geq \frac{1}{5}\|v\|_{1,*}^2, \qquad \forall v \in S_0(\Omega).$$
Owing to the Lax-Milgram Theorem [42], (2.1.8) admits a unique solution, so does (2.1.7). □

For three dimensional exterior domains Ω, we can define spaces $S(\Omega)$ and $S_0(\Omega)$ as we did for the two dimensional case.

Lemma 2.1.4. *If $u \in S(\Omega)$, $y_k = B_k u$, then there is a constant C independent of u, such that*
$$\frac{1}{C}\sum_{k=1}^{\infty} \xi^k |y_k - y_{k-1}|^2 \leq \int_{\Omega} |\nabla u|^2 \, dx \leq C\sum_{k=0}^{\infty} \xi^k |y_k|^2.$$

The proof of it is analogous to that of Lemma 2.1.2, thus omitted.

2.2 Transfer matrices

We consider the Laplace equation in this section. Firstly for the two dimensional problem we denote by

$$\begin{pmatrix} K_0 & -A^T \\ -A & K_0' \end{pmatrix}$$

the stiffness matrix on Ω_k as usual. This matrix is symmetric, because the bilinear functional $a(\cdot,\cdot)_{\Omega_k}$ is symmetric.

Lemma 2.2.1. K_0 and K_0' are positive definite matrices.

Proof. We take an arbitrary $y_0 \neq 0$ and set $y_1 = 0$. The interpolating functian u of them is not a constant, hence

$$W = \frac{1}{2}\int_{\Omega_1} |\nabla u|^2 \, dx > 0.$$

On the other hand

$$W = \frac{1}{2}(y_0^T \quad y_1^T)\begin{pmatrix} K_0 & -A^T \\ -A & K_0' \end{pmatrix}\begin{pmatrix} y_0 \\ y_1 \end{pmatrix} = \frac{1}{2}y_0^T K_0 y_0.$$

Thus

$$y_0^T K_0 y_0 > 0, \qquad \forall y_0 \neq 0,$$

so K_0 is a positive definite matrix. The argumant for K_0' is similar. □

By Lemma 2.1.3 the problem (2.1.7) admits a unique solution u for any $y_0 \in \mathbb{R}^n$ such that $B_0 u = y_0$. Let $y_k = B_k u$, then y_1 is determined by y_0 uniquely. It is easy to verify that the mapping $y_0 \to y_1$ is linear. Therefore there is a real matrix X such that

$$y_1 = X y_0.$$

Let $w(x) = u(\xi^k x)$, then $w \in S(\Omega)$, $B_0 w = y_k$ and w satisfies the equation (2.1.7), hence

$$y_{k+1} = X y_k.$$

We obtain by induction that

$$y_k = X^k y_0, \quad k = 1, 2, \cdots. \tag{2.2.1}$$

We will make use of the complex function solutions to the Laplace equation in this section. The above argument can be applied for this case without any modification except we regard that $y_1, y_2, \ldots, y_k, \ldots$ are vectors over the field of complex numbers. We observe that the transfer matrix X is real as we have proved.

2.2 TRANSFER MATRICES

Lemma 2.2.2. *If λ is an eigenvalue of the transfer matrix X and $|\lambda| \geq 1$, then $\lambda = 1$.*

Proof. Let g be the eigenvector associated with λ. Taking $y_0 = g$ as the boundary value of problem (2.1.7), we get $y_1, y_2, \ldots, y_k, \ldots$, and by (2.2.1)

$$y_k = X^k g = \lambda^k g.$$

Hence

$$\sum_{k=1}^{\infty} |y_k - y_{k-1}|^2 = \sum_{k=1}^{\infty} |\lambda^k - \lambda^{k-1}|^2 |g|^2 = \sum_{k=1}^{\infty} |\lambda|^{2k-2} |\lambda - 1|^2 |g|^2.$$

By Lemma 2.1.2, the series on the right hand side converges. We have $|g| \neq 0$, hence $\lambda = 1$. □

Lemma 2.2.3. *The elementary divisor corresponding to an eigenvalue $\lambda = 1$ of the transfer matrix X is linear.*

Proof. Suppose it were not the case, then there would be nonvanishing vectors g and g_0 such that

$$(X - I)g_0 = 0, \quad (X - I)g = g_0,$$

where I is a unit matrix, i.e.

$$Xg = g_0 + g.$$

By induction we would get

$$X^k g = k g_0 + g.$$

Taking $y_0 = g$ as the boundary value of the problem (2.1.7), we obtain $y_1, y_2, \ldots, y_k, \ldots$, then by

$$y_k = k g_0 + g$$

we would get

$$\sum_{k=1}^{\infty} |y_k - y_{k-1}|^2 = \sum_{k=1}^{\infty} |g_0|^2 = +\infty,$$

which contradicts $u \in H^{1,*}(\Omega)$ according to Lemma 2.1.2. □

Lemma 2.2.4. *There is an eigenvalue $\lambda = 1$ of the transfer matrix X, and the associated eigenvector is $g_1 = (1, 1, \ldots, 1)^T$.*

Proof. We take $y_0 = g_1$ as the boundary value of the problem (2.1.7) and get a constant solution $u \equiv 1$. Let $B_k u = y_k$, then $y_k = g_1$. By (2.2.1) we have

$$X g_1 = g_1.$$

□

Lemma 2.2.5. $X^k y_0 \to y_\infty$ $(k \to \infty)$ for any complex vector y_0, where

$$y_\infty = \alpha g_1,$$

and α is a constant.

Proof. On the basis of Lemma 2.2.2 and Lemma 2.2.3, we conclude that the limit of X^k exists as $k \to \infty$. Denote by y_∞ the limit of $X^k y_0$, which is $(y_\infty^{(1)}, y_\infty^{(2)}, \ldots, y_\infty^{(n)})^T$ in component form.

It will suffice to prove $y_\infty^{(1)} = y_\infty^{(2)} = \cdots = y_\infty^{(n)}$. If it were not true, we might assume that $y_\infty^{(i)} \neq y_\infty^{(i+1)}$, which correspond to the i-th node and the $(i+1)$-th node respectively. There must be one element e_j in Ω_k such that the connecting line segment of the above two nodes is one side of e_j. Let the length of this side be L, then we have

$$\int_{e_j} |\nabla u|^2 \, dx \geq \frac{\text{meas } e_j}{L^2} |y_k^{(i)} - y_k^{(i+1)}|^2.$$

By similarity $\frac{\text{meas } e_j}{L^2}$ is independent of k. Summing up the above inequality with respect to k, we obtain

$$\int_\Omega |\nabla u|^2 \, dx \geq \frac{\text{meas } e_j}{L^2} \sum_{k=1}^\infty |y_k^{(i)} - y_k^{(i+1)}|^2.$$

The general term of the series on the right hand side would take $|y_\infty^{(i)} - y_\infty^{(i+1)}|^2 \neq 0$ as its limit as $k \to \infty$, hence

$$\int_\Omega |\nabla u|^2 \, dx = \infty,$$

which leads to a contradiction. □

Lemma 2.2.6. *There is a unique eigenvalue $\lambda = 1$ of the transfer matrix X, and the absolute values of all other eigenvalues are less than 1.*

Proof. If it were not true, then by Lemma 2.2.2, Lemma 2.2.3 and Lemma 2.2.4, there would be another eigenvector g' associated with the eigenvalue $\lambda = 1$. According to Lemma 2.2.5 we have $X^k g' \to \alpha g_1$ $(k \to \infty)$, but $X^k g' = g'$, hence $g' = \alpha g_1$, which contradicts the fact that g_1 and g' are linearly independent. □

For the three dimensional problems we can also prove that K_0 and K_0' are symmetric and positive definite matrices. Moreover we have

Lemma 2.2.7. *Each eigenvalue λ of the transfer matrix X satisifies $|\lambda| < \xi^{-\frac{1}{2}}$.*

Proof. Let g be the eigenvector associated with λ. We take $y_0 = g$ as the boundary value of the problem (2.1.7) (three dimensional case), and get $y_1, y_2, \ldots, y_k, \ldots$, then $y_k = \lambda^k g$. Hence

$$\sum_{k=1}^\infty \xi^k |y_k - y_{k-1}|^2 = \sum_{k=1}^\infty \xi^k |\lambda|^{2k-2} |\lambda - 1|^2 |g|^2.$$

By Lemma 2.1.4 the above series converges. If $\lambda \neq 1$, then $\xi^k |\lambda|^{2k} < 1$, that is $|\lambda| < \xi^{-\frac{1}{2}}$, If $\lambda = 1$, then $y_0 = y_1 = \cdots = y_k = \cdots = g$. We have

$$\int_{\Omega_k} \frac{u^2(x)}{|x|^2} \, dx \geq C\xi^k,$$

thus $|u|_{1,*} = \infty$, which is unsuited to present needs. Therefore $\lambda = 1$ is not an eigenvalue. □

2.3 Further discussion for the infinite element spaces and the transfer matrices

We start with two dimensional problems. Let $\varphi_1, \varphi_2, \ldots \varphi_i, \cdots \in S_0(\Omega)$, φ_i is equal to 1 at a certain node of $\bigcup_{k=1}^{\infty} \Gamma_k$ and vanishes at other nodes, which can be named a "shape function". The set $\tilde{S}_0(\Omega)$ consists of all finite linear combinations

$$\varphi = c_1 \varphi_1 + c_2 \varphi_2 + \cdots + c_i \varphi_i$$

of $\varphi_i's$, then $\tilde{S}_0(\Omega) \subset S_0(\Omega)$. Each function in $\tilde{S}_0(\Omega)$ has a bounded support.

Theorem 2.3.1. $\tilde{S}_0(\Omega)$ is dense in $S_0(\Omega)$.

Proof. We take an arbitrary $u \in S_0(\Omega)$. Since $S_0(\Omega) \subset H_0^{1,*}(\Omega)$, there exists $u_1 \in C_0^{\infty}(\Omega)$, such that $\|u - u_1\|_{1,*} < \varepsilon$ for any $\varepsilon > 0$. We consider the following problem: find $u_2 \in S_0(\Omega)$, such that

$$a(u_2, v) = a(u_1, v), \qquad \forall v \in S_0(\Omega), \tag{2.3.1}$$

where $a(\cdot, \cdot)$ is defined by (2.1.6). We can prove that the problem (2.3.1) admits a unique solution as we did for the conclusion of Lemma 2.1.3. By the positive definiteness of the following quadratic functional we obtain

$$\frac{1}{2} a(u_2, u_2) - a(u_1, u_2) = \min_{v \in S_0(\Omega)} \left\{ \frac{1}{2} a(v, v) - a(u_1, v) \right\},$$

hence

$$\frac{1}{2} a(u_2, u_2) - a(u_1, u_2) \leq \frac{1}{2} a(u, u) - a(u_1, u).$$

By adding $\frac{1}{2} a(u_1, u_1)$ to the both sides of the above inequality we obtain

$$a(u_2 - u_1, u_2 - u_1) \leq a(u - u_1, u - u_1),$$

thus

$$a(u_2 - u_1, u_2 - u_1) < \varepsilon.$$

Applying the inequality (2.1.2) we get
$$\|u_2 - u_1\|_{1,*} < 5\varepsilon.$$
But $u_1 \in C_0^\infty(\Omega)$, so there exists a natural number k_0 such that $u_1 \equiv 0$ on the domain $\xi^{k_0}\Omega$. (2.1.6) and (2.3.1) imply that
$$a(u_2, v) = 0, \qquad \forall v \in S_0(\xi^{k_0}\Omega),$$
where we accept that every function in $S_0(\xi^{k_0}\Omega)$ is extended to zero on $\Omega \setminus \xi^{k_0}\Omega$. According to the conclusion of Lemma 2.2.5, u_2 tends to a constant α as $|x| \to \infty$, therefore u_2 is a bounded function.

We construct truncated functions $u_{2,k} \in S_0(\Omega), k = 1, 2, \ldots$ such that
$$u_{2,k} = \begin{cases} u_2, & \text{for } x \in \Omega \setminus \xi^k\Omega, \\ 0, & \text{for } x \in \xi^{k+1}\Omega. \end{cases}$$
Let $y_k = B_k u_2$, then by Lemma 2.1.2 we have
$$\int_\Omega |\nabla u_{2,k}|^2 \, dx \leq \int_\Omega |\nabla u_2|^2 \, dx + \frac{1}{C}|y_k|^2.$$
Therefore $\{u_{2,k}\}$ is bounded in $S_0(\Omega)$. Taking any $v \in S_0(\Omega)$ and making use of the Schwarz inequality we obtain
$$\left| \int_\Omega \nabla v \cdot \nabla (u_2 - u_{2,k}) \, dx \right|$$
$$= \left| \int_{\Omega_{k+1}} \nabla v \cdot \nabla (u_2 - u_{2,k}) \, dx + \int_{\xi^{k+1}\Omega} \nabla v \cdot \nabla u_2 \, dx \right|$$
$$\leq \left(\int_{\Omega_{k+1}} |\nabla v|^2 \, dx \right)^{\frac{1}{2}} \left(\int_{\Omega_{k+1}} |\nabla (u_2 - u_{2,k})|^2 \, dx \right)^{\frac{1}{2}} + \left| \int_{\xi^{k+1}\Omega} \nabla v \cdot \nabla u_2 \, dx \right|.$$
Because both u_2 and $u_{2,k}$ are bounded functions, $\int_{\Omega_{k+1}} |\nabla (u_2 - u_{2,k})|^2 \, dx$ are uniformly bounded with respect to k. And on account of $v, u_2 \in S_0(\Omega)$,
$$\int_{\Omega_{k+1}} |\nabla v|^2 \, dx \to 0, \quad \left| \int_{\xi^{k+1}\Omega} \nabla v \cdot \nabla u_2 \, dx \right| \to 0$$
as $k \to \infty$, therefore
$$\int_\Omega \nabla v \cdot \nabla (u_2 - u_{2,k}) \, dx \to 0$$
as $k \to \infty$, which means $u_{2,k}$ weakly converges to u_2. By the Mazur Theorem [62], there is a linear combination,
$$u_3 = c_1 u_{2,1} + c_2 u_{2,2} + \cdots + c_i u_{2,i},$$
such that $\|u_3 - u_2\|_{1,*} < \varepsilon$. Thus $\|u - u_3\|_{1,*} < 7\varepsilon$. But $u_3 \in \tilde{S}_0(\Omega)$, which leads to the conclusion. $\square$

Now we continue to discuss the Laplace equation. We give another formulation for the infinite element method: find $u \in S(\Omega)$, such that $u|_{\partial\Omega} = f$ and
$$a(u, \varphi_i) = 0, \qquad i = 1, 2, \cdots. \tag{2.3.2}$$

Theorem 2.3.2. *The formulation (2.3.2) is equivalent to (2.1.7).*

Proof. It will suffice to prove that (2.3.2) implies (2.1.7). Let u be the solution to (2.3.2), then u satisfies

$$a(u,v) = 0, \qquad \forall v \in \tilde{S}_0(\Omega),$$

which yields (2.1.7) by noting Theorem 2.3.1. □

Expressing (2.3.2) in terms of matrices, we get the system of equations in §1.1,

$$K_0 y_0 - A^T y_1 = f_0, \qquad (2.3.3)_0$$

$$-Ay_0 + K y_1 - A^T y_2 = 0, \qquad (2.3.3)_1$$

$$\cdots\cdots\cdots\cdots$$

$$-A y_{k-1} + K y_k - A^T y_{k+1} = 0, \qquad (2.3.3)_k$$

$$\cdots\cdots\cdots\cdots$$

and the equation for the transfer matrix X,

$$A^T X^2 - KX + A = 0. \qquad (2.3.4)$$

We should bear in mind that the solutions to (2.3.4) are not unique.

Theorem 2.3.3. *The necessary and sufficient conditions to determine the transfer matrix X are*

(a) *X satisfies (2.3.4);*

(b) *X admits an eigenvalue $\lambda = 1$ which corresponds to a linear elementary divisor and the eigenvector g_1;*

(c) *The absolute values of all other eigenvalues of X are less than 1.*

Proof. The lemmas from Lemma 2.2.2 to Lemma 2.2.6 and the equation (2.3.4) show that those conditions are necessary. It remains to prove they are sufficient. If a matrix X satisfies (a),(b),(c), let us prove it is the transfer matrix. We take an arbitrary $g \in \mathbb{R}^n$, and set $y_k = X^k g$, $k = 0, 1, \ldots$, then y_k satisfy $(2.3.3)_1, (2.3.3)_2, \cdots$. Let u be the interpolating function. If we can prove $u \in S(\Omega)$, then owing to Theorem 2.3.2, u is the solution to the problem (2.1.7), which means X is the desired one.

By (b) and (c), the limit of $X^k y_0$ exists as $k \to \infty$. Let it be y_∞. We have

$$X y_\infty = X \lim_{k \to \infty} X^k y_0 = \lim_{k \to \infty} X^{k+1} y_0 = y_\infty,$$

therefore $y_\infty = \alpha g_1$. Let $y'_k = y_k - y_\infty$, then the interpolating function of y'_k is just $u' = u - \alpha$. We only need to show $u' \in S(\Omega)$.

We express X in the Jordan canonical form,

$$X = TJT^{-1}, \qquad (2.3.5)$$

where
$$J = \begin{pmatrix} 1 & \\ & J_1 \end{pmatrix},$$

where J_1 consists of those Jordan blocks, the absolute values of the eigenvalues corresponding to which are less than 1. By the definition of y_k,

$$Xy'_{k-1} = X(y_{k-1} - y_\infty) = y_k - y_\infty = y'_k. \qquad (2.3.6)$$

Let
$$z_k = T^{-1} y'_k,$$

then (2.3.5) and (2.3.6) yield

$$z_k = J z_{k-1}, \qquad k = 1, 2, \cdots.$$

By induction we obtain
$$z_k = J^k z_0,$$

that is
$$z_k = \begin{pmatrix} 1 & \\ & J_1^k \end{pmatrix} z_0. \qquad (2.3.7)$$

Because $y'_k \to 0$ as $k \to \infty$, we have $z_k \to 0$. (2.3.7) implies that the first component of z_0 is zero, otherwise the limit of z_k would not be zero. Thus

$$z_k = \begin{pmatrix} 0 & \\ & J_1^k \end{pmatrix} z_0.$$

Let $\|\cdot\|$ be the spectral norm of a matrix. Denote by $\rho_1 = \rho(J_1)$ the spectral radius of the matrix J_1, and p the highest order among the Jordan blocks of J_1 whose spectral radius are equal to ρ_1, then [45]

$$\|J_1^k\| \le c(p) C_k^{p-1} \rho_1^{k-p+1}, \qquad (2.3.8)$$

where $k \ge p-1$, $c(p)$ is a constant depending only on p, and C_k^{p-1} is the combinational number. We have

$$\sum_{k=p-1}^{\infty} |z_k|^2 \le \sum_{k=p-1}^{\infty} \|J_1^k\|^2 |z_0|^2 \le \sum_{k=p-1}^{\infty} c^2(p) \left\{ C_k^{p-1} \rho_1^{k-p+1} \right\}^2 |z_0|^2.$$

It is known that $\rho_1 < 1$, hence the series on the right hand side converges. Therefore

$$\sum_{k=0}^{\infty} |y'_k|^2 = \sum_{k=0}^{\infty} |T z_k|^2 \le \|T\|^2 \sum_{k=0}^{\infty} |z_k|^2$$

converges too. On account of Lemma 2.1.2

$$\int_\Omega |\nabla u'|^2\, dx < +\infty.$$

Applying Lemma 2.1.1 we have $u' \in H^{1,*}(\Omega)$. □

For the corner problems of the Laplace equation, if the local boundary condition is $\frac{\partial u}{\partial \nu} = 0$, then all conclusions in the sections from §2.1 to §2.3 are valid. The basic spaces are $H^1(\Omega)$ and $H^1_{\Gamma_0}(\Omega)$, where

$$H^1_{\Gamma_0}(\Omega) = \{u \in H^1(\Omega); u|_{\Gamma_0} = 0\},$$

and the infinite element spaces are

$$S(\Omega) = \{u \in H^1(\Omega); u|_{e_i} \in P_1(e_i), i = 1, 2, \dots\},$$

$$S_0(\Omega) = \{u \in S(\Omega); u|_{\Gamma_0} = 0\}.$$

If we consider other boundary conditions, then the conclusions may change. We will discuss various boundary value problems of the system of the governing equations of plane elasticity in the next section. Since the argument is analogous, we will not expound the details for the Laplace equation here. But we would like to emphasize one point here, namely, Theorem 2.3.1 and its counterpart of the corner problems do not touch upon the equation at all, therefore they are generally applicable.

For the three dimensional exterior problems of the Laplace equation we have

Theorem 2.3.4. *The necessary and sufficient conditions to determine the transfer matrix X are*
(a) *X satisfies the equation*

$$\xi A^T X^2 - \xi^{\frac{1}{2}} KX + A = 0;$$

(b) *The absolute values of all eigenvalues of X are less than $\xi^{-\frac{1}{2}}$.*

2.4 Transfer matrices for the plane elasticity problems

The displacement vectors of the plane elasticity problems are denoted by $u = (u_1, u_2)$ or $v = (v_1, v_2)$. Let

$$a(u,v) = \int_\Omega \left\{ \left((\lambda + 2\mu)\frac{\partial u_1}{\partial x_1} + \lambda \frac{\partial u_2}{\partial x_2}\right)\frac{\partial v_1}{\partial x_1} + \left((\lambda + 2\mu)\frac{\partial u_2}{\partial x_2}\right.\right.$$
$$\left.\left. + \lambda \frac{\partial u_1}{\partial x_1}\right)\frac{\partial v_2}{\partial x_2} + \mu\left(\frac{\partial u_1}{\partial x_2} + \frac{\partial u_2}{\partial x_1}\right)\left(\frac{\partial v_1}{\partial x_2} + \frac{\partial v_2}{\partial x_1}\right)\right\} dx. \tag{2.4.1}$$

Sometimes the above bilinear functional (2.4.1) is denoted by $a(u,v)_\Omega$ to indicate the domain of integration. The definition for the infinite element space $S(\Omega)$ or $S_0(\Omega)$ changes with the problems, here are two examples,

(a) Exterior problems

$$S(\Omega) = \left\{ u \in (H^{1,*}(\Omega))^2; u\big|_{e_i} \in (P_1(e_i))^2, i = 1, 2, \ldots \right\},$$

$$S_0(\Omega) = \{ u \in S(\Omega); u\big|_{\partial\Omega} = 0\},$$

$$S(\partial\Omega) = \{ u\big|_{\partial\Omega}; u \in S(\Omega)\}.$$

(b) Corner problems

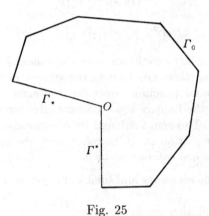

Fig. 25

For instance a fixed boundary condition is given on one adjacent side Γ^* to the corner point O, and a load free boundary condition is given on the other adjacent side Γ_* (Fig. 25), then

$$S(\Omega) = \{ u \in (H^1(\Omega))^2; u\big|_{e_i} \in (P_1(e_i))^2, i = 1, 2, \ldots, u\big|_{\Gamma^*} = 0\},$$

$$S_0(\Omega) = \{ u \in S(\Omega); u\big|_{\Gamma_0} = 0\},$$

$$S(\Gamma_0) = \{ u\big|_{\Gamma_0}; u \in S(\Omega)\}.$$

The setting of the infinite element method is: find $u \in S(\Omega)$ for any $f \in S(\Gamma_0)$ such that $u\big|_{\Gamma_0} = f$ and

$$a(u,v) = 0, \qquad \forall v \in S_0(\Omega). \tag{2.4.2}$$

In parallel with the propositions in §2.2 and §2.3, we establish the following lemmas and theorems. The proof of some propositions is the same as the previous one, thus omitted.

Lemma 2.4.1. K_0 and K_0' are positive definite matrices.

Proof. We take $y_0 \neq 0$ and $y_1 = 0$, then the corresponding interpolating function is not a rigid body motion, hence

$$W = \frac{1}{2}a(u,u) > 0.$$

It follows that K_0 is a positive definite matrix. The proof for K_0' is the same. □

Lemma 2.4.2. If λ is an eigenvalue of the transfer matrix X and $|\lambda| \geq 1$, then $\lambda = 1$.

Lemma 2.4.3. The elementary divisor associated with an eigenvalue $\lambda = 1$ of the transfer matrix X is linear.

Lemma 2.4.4. According to the local boundary condition, if the admissible translational motion is $u \equiv c$, where c is an arbitraty constant vector, then there are two eigenvalues $\lambda = 1$ of the transfer matrix X, and the associated eigenvectors are $g_1 = (1,0,1,0,\ldots,1,0)^T$ and $g_2 = (0,1,0,1,\ldots,0,1)^T$; if the admissible translational motion is $u_1 = c\cos\vartheta$, $u_2 = c\sin\vartheta$, where c is an arbitrary constant and ϑ is a fixed constant, then $c(g_1\cos\vartheta + g_2\sin\vartheta)$ are the only eigenvectors among the linear combinations of g_1 and g_2 associated with the eigenvalue $\lambda = 1$; if the admissible translational motion is $u \equiv 0$, then there is no eigenvector among the linear combinations of g_1 and g_2 associated with $\lambda = 1$.

Proof. Each linear combination of g_1 and g_2 is related to a translational motion. But any translational motion should satisfy the local boundary condition. Therefore only the above linear combinations of g_1 and g_2 are eigenvectors associated with the eigenvalue $\lambda = 1$. □

Lemma 2.4.5. $X^k y_0 \to y_\infty$ $(k \to \infty)$ for any complex vector y_0, where

$$y_\infty = \alpha g_1 + \beta g_2,$$

and α, β are constants.

Lemma 2.4.6. If an eigenvalue λ of the transfer matrix X is equal to 1, then the associated eigenvector of it is a linear combination of g_1 and g_2.

Theorem 2.4.1. The necessary and sufficient conditions to determine the transfer matrix X are

(a) X satisfies the equation,

$$A^T X^2 - KX + A = 0;$$

(b) If the admissible translational motion is $u \equiv c$, where c is an arbitrary constant vector, then there are two eigenvalues $\lambda = 1$ of X, the associated elementary divisors are linear, and g_1, g_2 are eigenvectors; if the admissible translational motion is $u_1 = c\cos\vartheta$, $u_2 = c\sin\vartheta$, where c is an arbitrary constant and ϑ is a fixed constant, then there is one eigenvalue $\lambda = 1$ of X, the associated elementary divisor is linear and $g_1 \cos \vartheta + g_2 \sin \vartheta$ is an eigenvector; if the admissible translational motion is $u \equiv 0$, then there is no eigenvalue of X equal to 1;

(c) The absolute values of all other eigenvalues of X are less than 1.

2.5 Combined stiffness matrices

In this section we deal with uniformly the various boundary value problems in the sections from §2.1 to §2.4. The transfer matrix X can always be expressed as

$$X = T \begin{pmatrix} I & \\ & J_1 \end{pmatrix} T^{-1},$$

where I is a unit matrix with order zero, one or two, which corresponds to constant solutions, and J_1 consists of the Jordan blocks associated with those eigenvalues whose absolute values are less than 1.

Lemma 2.5.1. *The eigenvectors of the transfer matrix X associated with the eigenvalue $\lambda = 1$ are necessarily the null eigenvectors of the combined stiffness matrix K_z.*

Proof. Let g be such an eigenvector. Since it corresponds to a constant solution, we have

$$W_1 = \frac{1}{2} \begin{pmatrix} g^T & g^T \end{pmatrix} \begin{pmatrix} K_0 & -A^T \\ -A & K_0' \end{pmatrix} \begin{pmatrix} g \\ g \end{pmatrix} = 0$$

on Ω_1. Here $\begin{pmatrix} K_0 & -A^T \\ -A & K_0' \end{pmatrix}$ is a symmetric and semi-positive definite matrix, therefore $\begin{pmatrix} g \\ g \end{pmatrix}$ is its null eigenvector, that is

$$\begin{pmatrix} K_0 & -A^T \\ -A & K_0' \end{pmatrix} \begin{pmatrix} g \\ g \end{pmatrix} = 0. \tag{2.5.1}$$

We obtain from the first row that

$$(K_0 - A^T)g = 0,$$

then by substituting $g = Xg$ into it we get

$$K_z g = (K_0 - A^T X)g = 0.$$

□

Lemma 2.5.2. If $u \in S(\Omega)$, $y_0 = B_0 u$ and u satisfies the equation

$$a(u,v) = 0, \qquad \forall v \in S_0(\Omega),$$

and if $w \in S(\Omega)$ and $z_0 = B_0 w$, then

$$a(u,w) = z_0^T K_z y_0.$$

Proof. We make a decomposition $w = w_1 + w_2$, where $w_1 \in S(\Omega)$, $w_1|_{\Gamma_k} = 0$, $k = 1, 2, \ldots$, and $w_2 \in S_0(\Omega)$, then

$$a(u, w_2) = 0.$$

Thus

$$\begin{aligned} a(u,w) &= a(u, w_1) \\ &= (\, z_0^T \quad 0\,) \begin{pmatrix} K_0 & -A^T \\ -A & K_0' \end{pmatrix} \begin{pmatrix} y_0 \\ X y_0 \end{pmatrix} \\ &= z_0^T K_z y_0. \end{aligned}$$

□

Theorem 2.5.1. K_z is a symmetric and semi-positive definite matrix.

Proof. We assume that the function w in Lemma 2.5.2 also satisfies

$$a(w,v) = 0, \qquad \forall v \in S_0(\Omega).$$

Since $a(\cdot, \cdot)$ is symmetric,

$$z_0^T K_z y_0 = a(u,w) = a(w,u) = y_0^T K_z z_0.$$

It implies that K_z is symmetric because y_0 and z_0 are arbitrary. Besides $a(\cdot, \cdot)$ is semi-positive definite, therefore

$$y_0^T K_z y_0 = a(u,u) \geq 0.$$

It implies that K_z is a semi-positive definite matrix because y_0 is arbitrary. □

2.6 Structure of the general solutions

An expression (1.1.13) for the transfer matrix was given in Chapter One, where it was assumed that all elementary divisors were linear. We will give a general expression for the transfer matrix in this section. Our procedure starts with analyzing the structure of the general solutions, the argument is effective for all boundary value problems in the previous sections. For definiteness we take the two dimensional exterior problem of the Laplace equation as an example.

Let Γ_0 be a polygon with the origin O in its interior. Γ_0 is star-shape with respect to the point O. Taking $\xi > 1$, we construct an infinite and similar partition as §1.1. Let the corresponding transfer matrix be X. We can also construct a similar partition concentrated on the point O infinitely with proportionality constants $\xi^{-1}, \xi^{-2}, \ldots, \xi^{-k}, \ldots$, and let the corresponding transfer matrix be $\tilde{X}$.

Denoting by Ω the exterior of Γ_0, and taking an arbitrary natural number $N \geq 2$, we consider the domain $\Omega \setminus \overline{\xi^N \Omega}$. If $y_0 \in \mathbb{C}^n$ and $y_N \in \mathbb{C}^n$ are known, then the finite element problem on this domain admits a unique solution, which satisfies the equation,

$$-Ay_{k-1} + Ky_k - A^T y_{k+1} = 0, \quad k = 1, 2, \ldots, N-1, \qquad (2.6.1)$$

or it is

$$\begin{pmatrix} K & -A \\ I & 0 \end{pmatrix} \begin{pmatrix} y_k \\ y_{k-1} \end{pmatrix} = \begin{pmatrix} A^T & 0 \\ 0 & I \end{pmatrix} \begin{pmatrix} y_{k+1} \\ y_k \end{pmatrix}$$

in the matrix form. If y_0 and y_N are arbitrary, then the space of solutions is $2n$-dimensional. By the matrix theory [43], there are two nonsingular matrices T_1 and T_2 such that

$$\begin{pmatrix} K & -A \\ I & 0 \end{pmatrix} = T_1 \Lambda_1 T_2, \quad \begin{pmatrix} A^T & 0 \\ 0 & I \end{pmatrix} = T_1 \Lambda_2 T_2, \qquad (2.6.2)$$

where Λ_1 and Λ_2 are block diagonal matrices with the same partition. Λ_1 may be expressed as

$$\Lambda_1 = \mathrm{diag}(J_1, J_2, J_3),$$

where J_1, J_2 and J_3 consist of the Jordan blocks associated with those eigenvalues satisfying $|\lambda| < 1$, $|\lambda| = 1$ and $1 < |\lambda| \leq \infty$ respectively. The submatrices in Λ_1 for $|\lambda| < \infty$ are the Jordan blocks in the ordinary sense, and the corresponding submatrices in Λ_2 are unit matrices, while the submatrices in Λ_1 for $\lambda = \infty$ are unit matrices and the corresponding submatrices in Λ_2 are backward shift Toeplitz matrices

$$\begin{pmatrix} 0 & 1 & & & \\ & \ddots & 1 & & \\ & & \ddots & \ddots & \\ & & & \ddots & 1 \\ & & & & 0 \end{pmatrix}$$

Now we investigate the structure of the matrix Λ_1.

Lemma 2.6.1. *The general solution to the system*

$$(K - A - A^T)z = 0 \qquad (2.6.3)$$

2.6 GENERAL SOLUTIONS

is $z = \alpha g_1$, where $g_1 = (1, 1, \ldots, 1)^T$ and α is an arbitrary constant.

Proof. By (2.5.1) we have

$$\begin{pmatrix} K_0 & -A^T \\ -A & K_0' \end{pmatrix} \begin{pmatrix} g_1 \\ g_1 \end{pmatrix} = 0.$$

Summing up the two rows, we get

$$(K - A - A^T)g_1 = 0,$$

hence g_1 is a solution. On the other hand, if z is a solution to (2.6.3), then it is easy to verify that

$$\begin{pmatrix} z^T & z^T \end{pmatrix} \begin{pmatrix} K_0 & -A^T \\ -A & K_0' \end{pmatrix} \begin{pmatrix} z \\ z \end{pmatrix} = 0.$$

Letting $y_0 = y_1 = z$, we construct the interpolating function u. The above equality yields $a(u, u)_{\Omega_1} = 0$. We obtain $u \equiv \alpha$. that is $z = \alpha g_1$. $\square$

Lemma 2.6.2. *The solution to the system*

$$(K - A - A^T)z = (A^T - A)g_1 \qquad (2.6.4)$$

exists.

Proof. Since

$$g_1^T(A^T - A)g_1 = 0,$$

Lemma 2.6.1 implies the existence of a solution. $\square$

Let

$$\varphi(\lambda) = \det(\lambda I - X),$$
$$f(\lambda) = \det \begin{pmatrix} K - \lambda A^T & -A \\ I & -\lambda I \end{pmatrix},$$

then we have

Lemma 2.6.3.

$$f(\lambda) = \lambda^n \varphi(\lambda) \varphi\left(\frac{1}{\lambda}\right) \det(A^T X - K).$$

Proof. Making use of the properties of determinants and the equation (2.3.4), we can deduce as follows:

$$f(\lambda) = \det \begin{pmatrix} K - \lambda A^T & \lambda(K - \lambda A^T) - A \\ I & 0 \end{pmatrix}$$
$$= \det(\lambda^2 A^T - \lambda K + A)$$
$$= \det((\lambda^2 A^T - \lambda K) - (A^T X^2 - KX))$$
$$= \det(A^T(\lambda^2 I - X^2) - K(\lambda I - X))$$
$$= \det((\lambda A^T + A^T X - K)(\lambda I - X))$$
$$= \varphi(\lambda) \det(\lambda A^T + A^T X - K).$$

On account of Theorem 2.5.1 we know $A^T X$ is a symmetric matrix, then we take the transpose of some matrices and obtain

$$f(\lambda) = \varphi(\lambda)\det(\lambda A + A^T X - K).$$

By noting the equation (2.3.4) we have

$$\begin{aligned}f(\lambda) &= \varphi(\lambda)\det(\lambda(KX - A^T X^2) + A^T X - K) \\ &= \varphi(\lambda)\det((A^T X - K)(I - \lambda X)) \\ &= \lambda^n \varphi(\lambda)\varphi\left(\frac{1}{\lambda}\right)\det(A^T X - K),\end{aligned}$$

if $\lambda \neq 0$. But both sides of the above expression are polynomials in λ, so it still holds for $\lambda = 0$. □

Using Lemma 2.6.3 and Theorem 2.3.3 we conclude that both J_1 and J_3 are $(n-1) \times (n-1)$ matrices, therefore J_2 is a 2×2 matrix. Now we go further into investigating the structure of the matrix J_2.

Lemma 2.6.4.
$$J_2 = \begin{pmatrix} 1 & 1 \\ 0 & 1 \end{pmatrix}.$$

Proof. By Lemma 2.6.1,

$$\begin{pmatrix} K - A^T & -A \\ I & -I \end{pmatrix}\begin{pmatrix} g_1 \\ g_1 \end{pmatrix} = 0,$$

hence $\lambda = 1$ is an eigenvalue. Applying Lemma 2.6.2 we get

$$\begin{pmatrix} K - A^T & -A \\ I & -I \end{pmatrix}\begin{pmatrix} z \\ z - g_1 \end{pmatrix} = \begin{pmatrix} A^T & 0 \\ 0 & I \end{pmatrix}\begin{pmatrix} g_1 \\ g_1 \end{pmatrix},$$

where z is the solution to the equation (2.6.4). Therefore the elementary divisor associated with $\lambda = 1$ is quadratic. □

We seek the general solution to the system (2.6.1) from the Jordan canonical form (2.6.2). Denote by $\varepsilon_1, \varepsilon_2, \ldots, \varepsilon_{2n}$ the standard basis of $\mathbb{C}^{2n}$, then we have

Theorem 2.6.1. *The general solutions to the system (2.6.1) are linear combinations of the following solutions:*
(a)
$$\begin{pmatrix} y_1 \\ y_0 \end{pmatrix} = T_2^{-1}\varepsilon_i, \quad \begin{pmatrix} y_2 \\ y_1 \end{pmatrix} = T_2^{-1}\Lambda_1\varepsilon_i, \ldots,$$
$$\begin{pmatrix} y_N \\ y_{N-1} \end{pmatrix} = T_2^{-1}\Lambda_1^{N-1}\varepsilon_i, \quad i = 1, \ldots, n+1,$$

(2.6.5)

(b)
$$\begin{pmatrix} y_N \\ y_{N-1} \end{pmatrix} = T_2^{-1}\varepsilon_i, \quad \begin{pmatrix} y_{N-1} \\ y_{N-2} \end{pmatrix} = T_2^{-1}\Lambda_2\varepsilon_i, \ldots,$$
$$\begin{pmatrix} y_1 \\ y_0 \end{pmatrix} = T_2^{-1}\Lambda_2^{N-1}\varepsilon_i, \quad i = n+2, \ldots, 2n. \tag{2.6.6}$$

Proof. For $i \leq n+1$, let

$$\begin{pmatrix} y_k \\ y_{k-1} \end{pmatrix} = T_2^{-1}\Lambda_1^{k-1}\varepsilon_i, \quad \begin{pmatrix} y_{k+1} \\ y_k^* \end{pmatrix} = T_2^{-1}\Lambda_1^k\varepsilon_i. \tag{2.6.7}$$

By (2.6.2) we get

$$\begin{pmatrix} K & -A \\ I & 0 \end{pmatrix} \begin{pmatrix} y_k \\ y_{k-1} \end{pmatrix} = T_1\Lambda_1^k\varepsilon_i,$$

$$\begin{pmatrix} A^T & 0 \\ 0 & I \end{pmatrix} \begin{pmatrix} y_{k+1} \\ y_k^* \end{pmatrix} = T_1\Lambda_2\Lambda_1^k\varepsilon_i.$$

The submatrices of Λ_2 corresponding to ε_i are unit matrices, therefore

$$\Lambda_2\Lambda_1^k\varepsilon_i = \Lambda_1^k\varepsilon_i.$$

Hence

$$\begin{pmatrix} K & -A \\ I & 0 \end{pmatrix} \begin{pmatrix} y_k \\ y_{k-1} \end{pmatrix} = \begin{pmatrix} A^T & 0 \\ 0 & I \end{pmatrix} \begin{pmatrix} y_{k+1} \\ y_k^* \end{pmatrix}.$$

From that we obtain $y_k = y_k^*$ and make sure that (2.6.5) indeed gives a set of solutions to the system (2.6.1). The same is true for $i \geq n+2$.

The columns of T_2^{-1} are linearly independent, hence (a) gives $(n+1)$ linearly independent solutions, and all vectors $\begin{pmatrix} y_{k+1} \\ y_k \end{pmatrix}$ are linear combinations of the first $(n+1)$ columns of T_2^{-1}. By the same reason, (b) gives $(n-1)$ linearly independent solutions and all vectors $\begin{pmatrix} y_{k+1} \\ y_k \end{pmatrix}$ are the linear combinations of the back $(n-1)$ columns of T_2^{-1}. Let us prove that the solutions in (a) and the solutions in (b) are linearly independent. Denote by E_a the set of the linear combinations of the solutions in (a), and by E_b the counterpart in (b), then they are $(n+1)$-dimensional and $(n-1)$-dimensional spaces respectively. We assume that there are nonvanishing elements $(y_0, y_1, \ldots, y_N) \in E_a$ and $(y_0^*, y_1^*, \ldots, y_N^*) \in E_b$ such that

$$c_1 y_k + c_2 y_k^* = 0, \quad k = 0, \ldots, N.$$

Taking $k = 0, 1$, we obtain

$$c_1 \begin{pmatrix} y_1 \\ y_0 \end{pmatrix} + c_2 \begin{pmatrix} y_1^* \\ y_0^* \end{pmatrix} = 0.$$

But $\begin{pmatrix} y_1 \\ y_0 \end{pmatrix}$ is a linear combination of the front $(n+1)$ columns of T_2^{-1} and does not vanish, and $\begin{pmatrix} y_1^* \\ y_0^* \end{pmatrix}$ is a linear combination of the back $(n-1)$ columns of T_2^{-1}. Consequently $c_1 = 0$ is the only possibility. By $c_2 y_k^* = 0$ we get $c_2 = 0$. Therefore any nonvanishing element in E_a is linearly independent of any nonvanishing element in E_b. In brief, we have already found $2n$ linearly independent solutions to the system (2.6.1). □

Making use of Lemma 2.2.4 and Lemma 2.2.5, we can decompose $y_0 = y_0' + \alpha g_1$ for any $y_0 \in \mathbb{C}^n$ such that $X^k y_0' \to 0$ $(k \to \infty)$. The set of y_0' forms a $(n-1)$-dimensional subspace of $\mathbb{C}^n$, which is denoted by E. We get another $(n-1)$-dimensional subspace $\tilde{E}$ of $\mathbb{C}^n$ from the matrix $\tilde{X}$ in much the same way.

Theorem 2.6.2. *Among the solutions listed in Theorem 2.6.1, $y_k = X^k y_0$, $y_0 \in E$ for $i = 1, \ldots, n-1$; $y_k = \tilde{X}^{N-k} y_N$, $y_N \in \tilde{E}$ for $i = n+2, \ldots, 2n$; and there are two particular solutions, $y_k = g_1$ $(k = 0, \ldots, N)$ and $y_k = z + k g_1$ $(k = 0, \ldots, N)$, where z is the solution to the system (2.6.4), for $i = n, n+1$.*

Proof. We consider the case of $i = 1, \ldots, n-1$ first, we can continue to define y_k $(k \geq N)$ after the manner of (2.6.5). It should be noticed that all corresponding eigenvalues λ satisfy $|\lambda| < 1$. Using the argument of Theorem 2.3.3, we can show that $y_1, y_2, \ldots, y_k, \ldots$ correspond to the infinite element solutions to (2.1.7). Therefore $y_k = X^k y_0$. Because $X^k y_0 \to 0$ $(k \to \infty)$, we have $y_0 \in E$.

The proof for $i = n+2, \ldots, 2n$ is the same. And the assertion for $i = n, n+1$ is easily obtained from Lemma 2.6.4. □

Corollary. *If $y_0 \in E$ and $y_k = X^k y_0$, then y_k is the linear combination of the solutions for $i = 1, \ldots, n-1$ in (2.6.5).*

Proof. It suffice to notice that both dimensions are $(n-1)$. □

Having finished the above preparation, we are now in a position to give the expression of the transfer matrix X. The matrix T_2^{-1} is partitioned as

$$T_2^{-1} = \begin{pmatrix} T_{11} & T_{12} \\ T_{21} & T_{22} \end{pmatrix},$$

where T_{11} and T_{21} are $n \times (n-1)$ matrices, T_{12} and T_{22} are $n \times (n+1)$ matrices. We construct $n \times n$ matrices

$$V = (T_{11} \quad g_1), \quad U = (T_{21} \quad g_1).$$

Taking an arbitrary $y_0 \in \mathbb{C}^n$, we decompose it as follows:

$$y_0 = y_0' + \alpha g_1, \quad y_0' \in E, \tag{2.6.8}$$

then by the Corollary of Theorem 2.6.2, there is a $z \in \mathbb{C}^{2n}$, the back $(n+1)$ components of which vanish, such that

$$\begin{pmatrix} X y_0' \\ y_0' \end{pmatrix} = T_2^{-1} z = \begin{pmatrix} T_{11} & 0 \\ T_{21} & 0 \end{pmatrix} z.$$

Let a vector w consist of the front $(n-1)$ components of z, then

$$Xy_0' = T_{11}w, \qquad y_0' = T_{21}w.$$

(2.6.8) gives

$$Xy_0 = T_{11}w + \alpha X g_1 = T_{11}w + \alpha g_1,$$

and

$$y_0 = T_{21}w + \alpha g_1.$$

Let

$$\tilde{w} = \begin{pmatrix} w \\ \alpha \end{pmatrix},$$

then

$$Xy_0 = V\tilde{w}, \qquad y_0 = U\tilde{w}.$$

But y_0 is arbitrary, so U is nonsingular, thus

$$Xy_0 = VU^{-1}y_0.$$

We obtain the expression for the transfer matrix X,

$$X = VU^{-1},$$

since y_0 is arbitrary.

The method which starts with studying the general solution to obtain the combined stiffness matrices was initialed by Thatcher. In his pioneer work the general solutions were obtained if the block A of the stiffness matrix was diagonal. The total computer work was to solve the eigenvalue problem of a $n \times n$ matrix. Now we know that the Fourier method is much convenient for this case.

2.7 Block circular stiffness matrices

Hereafter the finite element spaces defined on a domain Ω' are denoted by $S(\Omega')$, that is

$$S(\Omega') = \{u \in H^1(\Omega'); u\big|_{e_i} \in P_1(e_i), e_i \subset \Omega'\}.$$

We apply polar coordinates (r, ϑ) in this section. Following the notations in §1.2, let the nodes on Γ_0 be $(r, 0), (r, \frac{2\pi}{n}), \ldots, (r, \frac{2(n-1)\pi}{n})$. If $u, v \in S(\Omega_1)$, $B_0 u = y_0$, $B_1 u = y_1$, $B_0 v = z_0$, $B_1 v = z_1$, then

$$a(u,v)_{\Omega_1} = \begin{pmatrix} z_0^T & z_1^T \end{pmatrix} \begin{pmatrix} K_0 & -A^T \\ -A & K_0' \end{pmatrix} \begin{pmatrix} y_0 \\ y_1 \end{pmatrix}.$$

We define a matrix
$$B = \begin{pmatrix} 0 & & & & 1 \\ & \ddots & & & 1 \\ & & \ddots & \ddots & \\ & & & \ddots & 1 \\ 1 & & & & 0 \end{pmatrix}.$$

And set $\tilde{y}_0 = By_0$, $\tilde{y}_1 = By_1$, $\tilde{z}_0 = Bz_0$, $\tilde{z}_1 = Bz_1$. If the coordinate axes rotate $\frac{2\pi}{n}$ radian along the positive direction, then a mesh like Fig.4 is invariable under this transform, so is the equation $\triangle u = 0$, therefore the stiffness matrix is also invariable. We have
$$a(u,v)_{\Omega_1} = (\tilde{z}_0^T \quad \tilde{z}_1^T) \begin{pmatrix} K_0 & -A^T \\ -A & K_0' \end{pmatrix} \begin{pmatrix} \tilde{y}_0 \\ \tilde{y}_1 \end{pmatrix}.$$

Taking $z_1 = y_1 = 0$ we obtain
$$z_0^T K_0 y_0 = \tilde{z}_0^T K_0 \tilde{y}_0. \qquad (2.7.1)$$

In component form we write
$$K_0 = \begin{pmatrix} k_{11} & \cdots & k_{1n} \\ \cdots & \cdots & \cdots \\ k_{n1} & \cdots & k_{nn} \end{pmatrix}, \qquad (2.7.2)$$

and set $k_{i0} = k_{in}$, $k_{0j} = k_{nj}$ for convenience. Letting $\varepsilon_1, \ldots, \varepsilon_n$ be the standard basis of $\mathbb{C}^n$, and taking $y_0 = \varepsilon_j$, $z_0 = \varepsilon_i$, by (2.7.1) and (2.7.2) we get
$$k_{ij} = k_{i-1,j-1}, \qquad 1 \leq i, j \leq n.$$

Consequently K_0 is a circular matrix. By the same reason A and K_0' are circular matrices too.

We turn now to the plane elasticity problem in §1.8. We also have (2.7.1) and (2.7.2), where k_{ij} are 2×2 submatrices and the matrix B is
$$B = \begin{pmatrix} 0 & & & & I \\ & \ddots & & & I \\ & & \ddots & \ddots & \\ & & & \ddots & I \\ I & & & & 0 \end{pmatrix},$$

where I is a unit matrix of second order. We write
$$y_0 = \begin{pmatrix} y_0^{(1)} \\ y_0^{(2)} \\ \vdots \\ y_0^{(n)} \end{pmatrix}, \qquad z_0 = \begin{pmatrix} z_0^{(1)} \\ z_0^{(2)} \\ \vdots \\ z_0^{(n)} \end{pmatrix},$$

where $y_0^{(1)}, y_0^{(2)}, \ldots, y_0^{(n)}, z_0^{(1)}, z_0^{(2)}, \ldots, z_0^{(n)}$ are all two dimensional column vectors. Taking $y_0^{(j)}, z_0^{(i)}$ to be arbitrary and the other components of y_0 and z_0 to be zero, we get from (2.7.1) and (2.7.2) that

$$z_0^{(i)T} k_{ij} y_0^{(j)} = z_0^{(i)T} k_{i-1,j-1} y_0^{(j)}.$$

Then we get $k_{ij} = k_{i-1,j-1}$ because $y_0^{(j)}$ and $z_0^{(i)}$ are arbitrary. Therefore K_0 is a block circular matrix. By the same reason A and K_0' are block circular matrices too.

We conclude this section with considering the eigenvalues and eigenvectors of the matrices $X^{(i)} (i = 1, 2, \ldots, n)$ in §1.8. We take $y_k = g_1 = (1, 0, 1, 0, \ldots, 1, 0)^T, k = 0, 1, \ldots$, then $z_k = \bar{F} y_k = (\sqrt{n}, 0, \ldots, 0)^T$. Making use of the property of the matrix $X^{(1)}$ we get

$$X^{(1)} \begin{pmatrix} \sqrt{n} \\ 0 \end{pmatrix} = \begin{pmatrix} \sqrt{n} \\ 0 \end{pmatrix},$$

hence the matrix $X^{(1)}$ possesses an eigenvalue 1, which corresponds to the eigenvector $(\sqrt{n}, 0)^T$. Using the same argument and the vector $g_2 = (0, 1, 0, 1, \ldots, 0, 1)^T$ we can get another couple of eigenvalue 1 and eigenvector $(0, \sqrt{n})^T$ of the matrix $X^{(1)}$. Consequently $X^{(1)} = I$.

Matrices X and Z are similar to each other, so they possess the same set of eigenvalues. Owing to Theorem 2.4.1, X possesses only two eigenvalues $\lambda = 1$ and all other eigenvalues λ satisfy $|\lambda| < 1$. Now we have already found two eigenvalues $\lambda = 1$, hence all eigenvalues λ of $X^{(i)}$ $(i = 2, \ldots, n)$ satisfy $|\lambda| < 1$.

2.8 Iterative method of the first type

In this section we deal with uniformly the various boundary value problems in the sections from §2.1 to §2.4. The transfer matrix X is written in the Jordan canonical form

$$X = T \begin{pmatrix} I & \\ & J_1 \end{pmatrix} T^{-1},$$

where I is a unit matrix with order zero, one, or two, and J_1 consists of the Jordan blocks associated with those eigenvalues whose absolute values are less than 1. Denote by $Q(\Omega)$ (or $Q(\Omega_k)$) such a subspace of $S(\Omega)$ (or $S(\Omega_k)$) that $u \in Q(\Omega)$ (or $u \in Q(\Omega_k)$) if and only if $B_0 u = B_1 u = \cdots = B_k U = \ldots$, and $B_0 u$ is the eigenvector of the matrix X associated with the eigenvalue $\lambda = 1$.

Taking an arbitrary natural number k, we construct a subspace of $S(\Omega)$,

$$D_k(\Omega) = \left\{ u \in S(\Omega); u \big|_{\xi^k \Omega} \in Q(\xi^k \Omega) \right\}.$$

Now we fix a $f \in S(\Gamma_0)$ and assume that $u \in S(\Omega)$, $u \big|_{\Gamma_0} = f$, and u satisfies the equation,

$$a(u, v) = 0, \qquad \forall v \in S_0(\Omega). \tag{2.8.1}$$

Simultaneously we consider the following problem: find $\bar{u}_k \in D_k(\Omega)$, such that $\bar{u}_k|_{\Gamma_0} = f$ and $\bar{u}_k$ satisfies

$$a(\bar{u}_k, \bar{u}_k) = \min_{\substack{v|_{\Gamma_0}=f \\ v \in D_k(\Omega)}} a(v,v). \qquad (2.8.2)$$

The problem (2.8.2) admits a unique solution. Let $B_0 \bar{u}_k = y_0$, then there is a symmetric and semi-positive definite matrix $\bar{K}_z^{(k)}$, such that

$$a(\bar{u}_k, \bar{u}_k) = y_0^T \bar{K}_z^{(k)} y_0.$$

Taking $k = 2^l$, we get the very iterative schemes (1.3.3)–(1.3.5),(1.3.7).

Theorem 2.8.1. *For the two dimensional problems we have*
(a) $K_z \leq \bar{K}_z^{(k+1)} \leq \bar{K}_z^{(k)}$,
(b) $\|\bar{K}_z^{(k)} - K_z\| \leq \left(c(p)(\kappa(T))^2 C_{k-1}^{p-1} \rho_1^{k-p}\right)^2 \|K_0 - K_z\|$,
for $k = p, p+1, \ldots$, where $\kappa(T) = \|T\| \|T^{-1}\|$ is the condition number of the matrix T, and the other notations are the same as that in (2.3.8).

Proof. (a) is valid because $D_k(\Omega) \subset D_{k+1}(\Omega) \subset S(\Omega)$.

It remains to prove (b). We decompose the vector y_0 into $y_0 = y_0^{(1)} + y_0^{(2)}$, where $y_0^{(1)}$ belongs to the invariant subspace of the transfer matrix X associated with $|\lambda| < 1$, and $y_0^{(2)}$ belongs to the one associated with $\lambda = 1$. Let

$$y_k^{(1)} = X^k y_0^{(1)}, \qquad y_k^{(2)} = X^k y_0^{(2)},$$

and denote by $u^{(1)}$ and $u^{(2)}$ the corresponding interpolating functions, then we have $u = u^{(1)} + u^{(2)}$, and $u^{(2)} \in Q(\Omega)$. Consequently

$$a(u,u) = a(u^{(1)}, u^{(1)}).$$

Let $v \in S(\Omega)$ satisfy

$$v = \begin{cases} u^{(1)}, & x \in \Omega \setminus \xi^{k-1}\Omega, \\ 0, & x \in \xi^k \Omega, \end{cases}$$

then $v + u^{(2)} \in D_k(\Omega)$, and $B_0(v + u^{(2)}) = y_0$, hence

$$a(\bar{u}_k, \bar{u}_k) \leq a(v + u^{(2)}, v + u^{(2)}) = a(v, v),$$

accordingly

$$\begin{aligned}
0 \leq y_0^T(\bar{K}_z^{(k)} - K_z)y_0 &= a(\bar{u}_k, \bar{u}_k) - a(u,u) \\
&\leq a(v,v) - a(u^{(1)}, u^{(1)}) \\
&= a(v,v)_{\Omega_k} - a(u^{(1)}, u^{(1)})_{\xi^{k-1}\Omega}.
\end{aligned}$$

Using the notation of the stiffness matrix of one layer, we can write
$$a(v,v)_{\Omega_k} = \begin{pmatrix} (y_{k-1}^{(1)})^T & 0 \end{pmatrix} \begin{pmatrix} K_0 & -A^T \\ -A & K_0' \end{pmatrix} \begin{pmatrix} y_{k-1}^{(1)} \\ 0 \end{pmatrix}$$
$$= (y_{k-1}^{(1)})^T K_0 y_{k-1}^{(1)}.$$

We also have
$$a(u^{(1)}, u^{(1)})_{\xi^{k-1}\Omega} = (y_{k-1}^{(1)})^T K_z y_{k-1}^{(1)}.$$

Therefore
$$0 \le y_0^T (\bar{K}_z^{(k)} - K_z) y_0 \le (y_{k-1}^{(1)})^T (K_0 - K_z) y_{k-1}^{(1)}. \tag{2.8.3}$$

It is not hard to show that
$$|y_0^{(1)}| \le \kappa(T)|y_0|.$$

By (2.3.8) we have
$$|y_{k-1}^{(1)}| = \left| T \begin{pmatrix} I & \\ & J_1^{k-1} \end{pmatrix} T^{-1} y_0^{(1)} \right|$$
$$\le c(p)(\kappa(T))^2 C_{k-1}^{p-1} \rho_1^{k-p} |y_0|.$$

Then substituting it into (2.8.3) we obtain
$$0 \le y_0^T (\bar{K}_z^{(k)} - K_z) y_0$$
$$\le \left(c(p)(\kappa(T))^2 C_{k-1}^{p-1} \rho_1^{k-p} \right)^2 \|K_0 - K_z\| \cdot |y_0|^2.$$

On account of the already proved conclusion (a), $\bar{K}_z^{(k)} - K_z$ is a symmetric and semi-positive definite matrix, therefore
$$\|\bar{K}_z^{(k)} - K_z\| \le \sup_{|y_0|=1} y_0^T (\bar{K}_z^{(k)} - K_z) y_0$$
$$\le \left(c(p)(\kappa(T))^2 C_{k-1}^{p-1} \rho_1^{k-p} \right)^2 \|K_0 - K_z\|.$$

□

Theorem 2.8.2. *For the three dimensional problems we have*
(a) $K_z \le \bar{K}_z^{(k+1)} \le \bar{K}_z^{(k)}$,
(b) $\|\bar{K}_z^{(k)} - K_z\| \le \left(c(p)(\kappa(T))^2 C_{k-1}^{p-1} \rho_1^{k-p} \right)^2 \xi^{k-1} \|K_0 - K_z\|$,
for $k = p, p+1, \cdots$.

Proof. We notice that
$$a(v,v)_{\Omega_k} = \xi^{k-1} \begin{pmatrix} (y_{k-1}^{(1)})^T & 0 \end{pmatrix} \begin{pmatrix} K_0 & -A^T \\ -A & K_0' \end{pmatrix} \begin{pmatrix} y_{k-1}^{(1)} \\ 0 \end{pmatrix},$$
$$a(u^{(1)}, u^{(1)})_{\xi^{k-1}\Omega} = \xi^{k-1} (y_{k-1}^{(1)})^T K_z y_{k-1}^{(1)},$$

by §1.5. Then the proof follows the same lines as those of Theorem 2.8.1   □

Remarks.

(a) The rate of convergence for the two dimensional problems is $O(\rho_1^{2^{l+1}})$, where $0 < |\rho_1| < 1$, therefore the rate of this iterative scheme is very high.

(b) The rate of convergence for the three dimensional problems is $O\left((\sqrt{\xi}\rho_1)^{2^{l+1}}\right)$. By Lemma 2.2.7, $0 < |\sqrt{\xi}\rho_1| < 1$, so this rate is also very high.

(c) If we can find a space $D_k^*(\Omega)$, such that $D_k(\Omega) \subset D_k^*(\Omega) \subset S(\Omega)$, then we can construct a solution $u_k^* \in D_k^*(\Omega)$, which satisfies $u_k^*|_{\Gamma_0} = f$ and

$$a(u_k^*, u_k^*) = \min_{\substack{v|_{\Gamma_0}=f \\ v \in D_k^*(\Omega)}} a(v,v).$$

Accordingly an approximate stiffness matrix K_z^* is obtained which satisfies $K_z \leq K_z^* \leq \bar{K}_z^{(k)}$, therefore K_z^* is closer to K_z than $\bar{K}_z^{(k)}$. This is the valid reason for the modified scheme applied in §1.7 and §1.8.

2.9 Iterative method of the second type

Uniformly in this section we deal with the various boundary value problems of the Laplace equation and the boundary value problems of the system of plane elasticity equations which are classified in the " + " category of Table 2 in §1.8.

We start with two dimensional problems. To give an exposition of the method, we first prove the following lemma.

Lemma 2.9.1. *The problem* (2.8.1) *is equivalent to:* find $u \in S(\Omega \setminus \overline{\xi^k \Omega})$, *such that* $u|_{\Gamma_0} = f$, *and*

$$a(u,v)_{\Omega \setminus \overline{\xi^k \Omega}} + y_k^T K_z z_k = 0, \quad \forall v \in S_0(\Omega \setminus \overline{\xi^k \Omega}), \tag{2.9.1}$$

where K_z is the combined stiffness matrix, $k \geq 1$, *and* $y_k = B_k u$, $z_k = B_k v$.

Proof. If u is the solution to (2.8.1) then by Lemma 2.5.2,

$$a(u,v) = a(u,v)_{\Omega \setminus \overline{\xi^k \Omega}} + y_k^T K_z z_k \tag{2.9.2}$$

for all $v \in S_0(\Omega)$. Therefore u is the solution to (2.9.1). Conversely, if u is the solution to (2.9.1), then we extend u to the domain Ω by setting $y_{k+l} = X^l y_k$. Again by Lemma 2.5.2, (2.9.2) holds for all $v \in S_0(\Omega)$, which means u is also a solution to (2.8.1). $\square$

Motivated by Lemma 2.9.1 we take an arbitrary symmetric and semi-positive definite matrix $K_z^{(0)}$, and construct a quadratic functional in $S(\Omega)$,

$$a_k(u,u) = a(u,u)_{\Omega \setminus \overline{\xi^k \Omega}} + y_k^T K_z^{(0)} y_k,$$

where $y_k = B_k u, k \geq 1$. Fix an arbitrary $f \in S(\Gamma_0)$, and let u_k be the solution, which is unique on the domain $\Omega \setminus \overline{\xi^k \Omega}$, to the following problem,

$$a_k(u_k, u_k) = \min_{\substack{v \in S(\Omega) \\ v|_{\Gamma_0} = f}} a_k(v, v). \tag{2.9.3}$$

There exists a symmetric and semi-positive definite matrix $K_z^{(k)}$, satisfying

$$a_k(u_k, u_k) = y_0^T K_z^{(k)} y_0,$$

where $y_0 = B_0 u_k$.

To prove $K_z^{(k)} \to K_z$ as $k \to \infty$, we consider an auxiliary problem: find $\tilde{u}_k \in S(\Omega \setminus \overline{\xi^k \Omega})$, such that $\tilde{u}_k|_{\Gamma_0} = f$ and $\tilde{u}_k$ satisfies

$$a(\tilde{u}_k, \tilde{u}_k)_{\Omega \setminus \overline{\xi^k \Omega}} = \min_{\substack{v \in S(\Omega \setminus \overline{\xi^k \Omega}) \\ v|_{\Gamma_0} = f}} a(v, v)_{\Omega \setminus \overline{\xi^k \Omega}}. \tag{2.9.4}$$

There exists a symmetric and semi-positive definite matrix $\tilde{K}_z^{(k)}$ satisfying

$$a(\tilde{u}_k, \tilde{u}_k)_{\Omega \setminus \overline{\xi^k \Omega}} = y_0^T \tilde{K}_z^{(k)} y_0,$$

where $y_0 = B_0 \tilde{u}_k$.

Lemma 2.9.2. Let $u, \tilde{u}_k$ be the solutions to (2.8.1) and (2.9.4) respectively, then we have

$$\lim_{k \to \infty} a(\tilde{u}_k, \tilde{u}_k)_{\Omega \setminus \overline{\xi^k \Omega}} = a(u, u)_\Omega, \tag{2.9.5}$$

that is

$$\lim_{k \to \infty} \tilde{K}_z^{(k)} = K_z.$$

Moreover we have

$$\tilde{K}_z^{(k)} \leq \tilde{K}_z^{(k+1)} \leq K_z. \tag{2.9.6}$$

Proof. It is evident that

$$\begin{aligned} a(\tilde{u}_k, \tilde{u}_k)_{\Omega \setminus \overline{\xi^k \Omega}} &\leq a(\tilde{u}_{k+1}, \tilde{u}_{k+1})_{\Omega \setminus \overline{\xi^k \Omega}} \\ &\leq a(\tilde{u}_{k+1}, \tilde{u}_{k+1})_{\Omega \setminus \overline{\xi^{k+1} \Omega}} \\ &\leq a(u, u)_{\Omega \setminus \overline{\xi^{k+1} \Omega}} \leq a(u, u)_\Omega, \end{aligned} \tag{2.9.7}$$

hence (2.9.6) holds. Now we fix a natural number l, then

$$a(\tilde{u}_k, \tilde{u}_k)_{\Omega \setminus \overline{\xi^l \Omega}} \leq a(\tilde{u}_k, \tilde{u}_k)_{\Omega \setminus \overline{\xi^k \Omega}} \tag{2.9.8}$$

for $k \geq l$. Combining (2.9.7) and (2.9.8) we get

$$a(\tilde{u}_k, \tilde{u}_k)_{\Omega \setminus \overline{\xi^l \Omega}} \leq a(u, u)_\Omega,$$

which implies that the left hand side are uniformly bounded with respect to k. Since $\tilde{u}_k|_{\Gamma_0} = f$ is fixed, by the Friedrichs inequality or the Korn inequality we conclude that $\|\tilde{u}_k\|_{1, \Omega \setminus \overline{\xi^l \Omega}}$ are uniformly bounded with respect to k.

$S(\Omega \setminus \overline{\xi^l \Omega})$ is a finite dimensional space, therefore all norms are equivalent. We take a subsequence which is especially convergent with respect to the norm of H^1, then take a diagonal subsequence with respect to l, and let $\tilde{u}$ be the limit. Thus $\tilde{u} \in S(\Omega \setminus \overline{\xi^l \Omega})$ for all natural number l, $\tilde{u}|_{\Gamma_0} = f$, and

$$a(\tilde{u}, \tilde{u})_{\Omega \setminus \overline{\xi^l \Omega}} \leq a(u, u)_\Omega.$$

Letting $l \to \infty$ we get

$$a(\tilde{u}, \tilde{u})_\Omega \leq a(u, u)_\Omega. \tag{2.9.9}$$

If Ω is a bounded domain, then an application of the Friedrichs inequality or the Korn inequality shows that $\tilde{u} \in H^1(\Omega)$. If Ω is an unbounded domain, then for the Laplace equation by Lemma 2.1.1 $\tilde{u} \in H^{1,*}(\Omega)$, that is $\tilde{u} \in S(\Omega)$. For the system of plane elasticity equations on unbounded domains, let us assume that the boundary condition is $u = 0$, then applying the Korn inequality we get

$$\int_{\Omega_k} |\nabla \tilde{u}|^2 \, dx \leq C a(\tilde{u}, \tilde{u})_{\Omega_k}.$$

The constant C is independent of k because of similarity. Summing them up with respect to k, we obtain

$$\int_\Omega |\nabla \tilde{u}|^2 \, dx \leq C a(\tilde{u}, \tilde{u})_\Omega,$$

then applying Lemma 2.1.1 we also get $\tilde{u} \in H^{1,*}(\Omega)$. Since u is the minimum point, (2.9.9) implies

$$a(\tilde{u}, \tilde{u})_\Omega = a(u, u)_\Omega.$$

By uniqueness of the solution we obtain $\tilde{u} = u$.

The original sequence $\{\tilde{u}_k\}$ converges to u because the limit function is unique. Now we have

$$\lim_{k \to \infty} a(\tilde{u}_k, \tilde{u}_k)_{\Omega \setminus \overline{\xi^l \Omega}} = a(u, u)_{\Omega \setminus \overline{\xi^l \Omega}}.$$

Then (2.9.7) and (2.9.8) yield (2.9.5). □

Theorem 2.9.1. *For two dimensional problems if the eigenvectors associated with the eigenvalue $\lambda = 1$ of the transfer matrix X are null eigenvectors of $K_z^{(0)}$, then*

$$\lim_{k \to \infty} K_z^{(k)} = K_z. \tag{2.9.10}$$

Proof. Take an arbitrary $f \in S(\Gamma_0)$. If $\bar{u}_k, u_k, \tilde{u}_k$ are the solutions to (2.8.2), (2.9.3),(2.9.4) respectively, and $\bar{K}_z^{(k)}$, $K_z^{(k)}$, $\tilde{K}_z^{(k)}$ are the corresponding combined stiffness matrices, then by Theorem 2.8.1, Theorem 2.8.2 and Lemma 2.9.2, we obtain

$$\tilde{K}_z^{(k)} \leq K_z \leq \bar{K}_z^{(k)},$$

and

$$\lim_{k \to \infty} \bar{K}_z^{(k)} = \lim_{k \to \infty} \tilde{K}_z^{(k)} = K_z. \tag{2.9.11}$$

Let $y_k = B_k u_k$, then

$$a_k(u_k, u_k) = a(u_k, u_k)_{\Omega \setminus \overline{\xi^k \Omega}} + y_k^T K_z^{(0)} y_k.$$

Since $K_z^{(0)}$ is a semi-positive definite matrix,

$$a_k(u_k, u_k) \geq a(u_k, u_k)_{\Omega \setminus \overline{\xi^k \Omega}} \geq a(\tilde{u}_k, \tilde{u}_k)_{\Omega \setminus \overline{\xi^k \Omega}}.$$

By the hypothesis of this theorem

$$K_z^{(0)} B_k \bar{u}_k = 0.$$

Therefore

$$a_k(u_k, u_k) \leq a_k(\bar{u}_k, \bar{u}_k) = a_k(\bar{u}_k, \bar{u}_k)_{\Omega \setminus \overline{\xi^k \Omega}}.$$

Combining the above inequalities we have

$$a(\tilde{u}_k, \tilde{u}_k)_{\Omega \setminus \overline{\xi^k \Omega}} \leq a_k(u_k, u_k) \leq a(\bar{u}_k, \bar{u}_k)_\Omega,$$

that is

$$\tilde{K}_z^{(k)} \leq K_z^{(k)} \leq \bar{K}_z^{(k)}.$$

Then (2.9.11) gives (2.9.10). □

We turn now to the three dimensional exterior problems of the Laplace equation. We construct a quadratic functional in $S(\Omega)$,

$$a_k(u, u) = a(u, u)_{\Omega \setminus \overline{\xi^k \Omega}} + \xi^k y_k^T K_z^{(0)} y_k, \tag{2.9.12}$$

where $y_k = B_k u, k \geq 1$, and $K_z^{(0)}$ is a given symmetric and positive definite matrix. Fixing an arbitrary $f \in S(\Gamma_0)$, let u_k be the solution to the problem,

$$a_k(u_k, u_k) = \min_{\substack{v \in S(\Omega) \\ v|_{\Gamma_0} = f}} a_k(v, v). \tag{2.9.13}$$

Analogously there exists a symmetric and positive definite matrix $K_z^{(k)}$, satisfying

$$a_k(u_k, u_k) = y_0^T K_z^{(k)} y_0,$$

where $y_0 = B_0 u_k$.

Theorem 2.9.2. *For the three dimensional exterior problems of the Laplace equation, we have*
$$\lim_{k \to \infty} K_z^{(k)} = K_z.$$

Proof. We apply the results in the previous section. Let
$$D_k(\Omega) = \left\{ u \in S(\Omega); u\big|_{\xi^k \Omega} = 0 \right\},$$

and $\bar{u}_k$ be the solution to (2.8.2), then (2.9.12) and (2.9.13) give
$$a_k(u_k, u_k) \leq a_k(\bar{u}_k, \bar{u}_k) = a(\bar{u}_k, \bar{u}_k)_{\Omega \setminus \overline{\xi^k \Omega}}. \tag{2.9.14}$$

By Theorem 2.8.2, the right hand side decreases as $k \to \infty$, thus it is bounded, therefore
$$a_k(u_k, u_k) \leq C.$$

Applying (2.9.12) we get
$$a(u_k, u_k)_{\Omega \setminus \overline{\xi^k \Omega}} \leq C, \quad \xi^k y_k^T K_z^{(0)} y_k \leq C.$$

Since $K_z^{(0)}$ is positive definite, we have
$$|y_k| \leq C \xi^{-\frac{k}{2}}.$$

We extend the function u_k such that $u_k \in S(\Omega)$ and $u_k\big|_{\xi^{k+1}\Omega} \equiv 0$. Using the expression of the stiffness matrix on Ω_{k+1} we get
$$a(u_k, u_k)_{\Omega_{k+1}} = \xi^k \begin{pmatrix} y_k^T & 0 \end{pmatrix} \begin{pmatrix} K_0 & -A^T \\ -A & K_0' \end{pmatrix} \begin{pmatrix} y_k \\ 0 \end{pmatrix} \leq C.$$

Consequently $a(u_k, u_k) \leq C$, then the inequality (2.1.1) implies $\|u_k\|_{1,*} \leq C$. We take a weakly convergent subsequence in $H^{1,*}(\Omega)$ and by analogy with Lemma 2.9.1 we further take a subsequence, still denoted by $\{u_k\}$, such that u_k strongly converges in $H^1(\Omega \setminus \overline{\xi^l \Omega})$ for all natural number l. Let $\tilde{u} \in H^{1,*}(\Omega)$ be the limit. Taking limit in the equations (2.3.2) shows that $\tilde{u}$ is the solution to (2.8.1). Uniqueness of the solution implies $\tilde{u} = u$.

We fix a natural number l, then for $k > l$
$$a_k(u_k, u_k) \geq a(u_k, u_k)_{\Omega \setminus \overline{\xi^l \Omega}} \to a(u, u)_{\Omega \setminus \overline{\xi^l \Omega}} \quad (k \to \infty).$$

Therefore
$$\lim_{k \to \infty} a_k(u_k, u_k) \geq a(u, u)_{\Omega \setminus \overline{\xi^l \Omega}}.$$

Since l is arbitrary, we have

$$\lim_{k\to\infty} a_k(u_k, u_k) \geq a(u,u)_\Omega.$$

On the other hand Theorem 2.8.2 and (2.9.14) yield

$$\overline{\lim_{k\to\infty}} \, a_k(u_k, u_k) \leq a(u,u)_\Omega.$$

Therefore

$$\lim_{k\to\infty} a_k(u_k, u_k) = a(u,u)_\Omega,$$

which means $K_z^{(k)} \to K_z$. By uniqueness of the combined stiffness matrix K_z, we get the convergence of the original sequence $\{K_z^{(k)}\}$. □

Now we apply Theorem 2.9.1 and Theorem 2.9.2 to investigate the equations satisfied by the combined stiffness matrix K_z. By (1.3.10), K_z satisfies

$$K_z = K_l - A_l^T (K_z + K_l')^{-1} A_l \tag{2.9.15}$$

for the two dimensional case, where l is an arbitrary natural number. And by §1.5 K_z satisfies

$$K_z = K_l - A_l^T (\xi^{2^l} K_z + K_l')^{-1} A_l \tag{2.9.16}$$

for the three dimensional case. For the two dimensional problems if $K_z^{(0)}$ is a $n \times n$ real symmetric and semi-positive definite matrix, and the eigenvectors associated with the eigenvalue $\lambda = 1$ of the transfer matrix X are null eigenvectors of $K_z^{(0)}$, then we denote $K_z^{(0)} \in M$. For the three dimensional problems let M be the set of all $n \times n$ real symmetric and positive definite matrices. We have the following theorem:

Theorem 2.9.3. *Equation (2.9.15) or (2.9.16) admits a unique solution in the set M.*

Proof. We have proved in Chapter One that the matrix K_z satisfies the equation (2.9.15) or (2.9.16). By Lemma 2.5.1 and Theorem 2.5.1, we have $K_z \in M$ for the two dimensional problems. For the three dimensional exterior problems of the Laplace equation, if $y_0^T K_z y_0 = 0$ for a certain y_0, then the solution u satisfies $\|\nabla u\|_{0,\Omega} = 0$. Hence u is a constant. But the constant must be zero, so $y_0 = 0$, which means K_z is a positive definite matrix. We also have $K_z \in M$. It remains to prove the uniqueness.

Let $K_z^{(0)} \in M$ be a solution to (2.9.15). Taking $k = 2^l$ in the equation (2.9.3) we get $K_z^{(k)} \in M$. $K_z^{(k)}$ satisfies the recursion formula,

$$K_z^{(k)} = K_l - A_l^T (K_z^{(0)} + K_l')^{-1} A_l.$$

Again we regard $K_z^{(k)}$ as $K_z^{(0)}$ and get $K_z^{(2k)}$ from (2.9.3), repeating this procedure we get a sequence $\{K_z^{(mk)}\}_{m=0}^\infty$. But $K_z^{(0)}$ is a solution to (2.9.15), so $K_z^{(0)} = K_z^{(k)} = \cdots = K_z^{(mk)} = \cdots$. Making use of Theorem 2.9.1 we get $K_z^{(0)} = K_z$. The proof for the equation (2.9.16) is the same. □

Based on Theorem 2.9.3 we convince ourselves that the limit of the iterative scheme of the second type is independent of the round-off error of the procedure.

2.10 General elliptic systems

It is possible to extend the results in the above sections to general $2m$-th order elliptic systems. Let the corresponding bilinear functional be

$$a(u,v) = \sum_{i,j=1}^{N} \sum_{|\alpha|,|\beta|=m} \int_\Omega a_{ij\alpha\beta} \partial^\alpha u_i \partial^\beta v_j \, dx, \qquad (2.10.1)$$

where $x \in \mathbb{R}^d$, Ω is an open set in $\mathbb{R}^d$, $u = (u_1, \ldots, u_N)$, $v = (v_1, \ldots, v_N)$, $u, v \in (H^m(\Omega))^N$, ∂^α and ∂^β are partial derivative operators, α and β are multi-indices, and $a_{ij\alpha\beta}$ are constants. The above bilinear functional is occasionally denoted by $a(u,v)_\Omega$ if to indicate the set Ω is necessary.

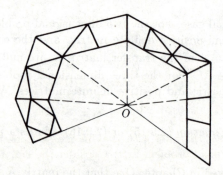

Fig. 26

Let Ω be a neighborhood of a corner O as shown in Fig.26, which is star-shape with respect to the point O, that is the line segment from each point in this domain to the point O lies entirely in Ω. Taking ξ, $0 < \xi < 1$, we define surfaces $\Gamma_0, \Gamma_1, \ldots, \Gamma_k, \ldots$ as we did in §1.7. The domain between Γ_k and the point O is denoted by $\xi^k \Omega$, then we set $\Omega_k = \xi^{k-1}\Omega \setminus \overline{\xi^k \Omega}$. The partition of each layer Ω_k is regarded as a "substructure". It is demanded that the meshes in all layers are geometrically similar to each other.

Denote by $S(\Omega_k)$ the finite dimensional subspace of $(H^m(\Omega_k))^N$ corresponding to the above partition. If $u \in S(\Omega_k)$, then its traces from order zero to $(m-1)$ on Γ_{k-1}

2.10 GENERAL ELLIPTIC SYSTEMS

and Γ_k make sense [60]. We denote the trace operators by γ_{k-1} and γ_k, and the trace spaces by $S(\Gamma_{k-1})$ and $S(\Gamma_k)$. Define $S_0(\Omega_k) = \{u \in S(\Omega_k); \gamma_{k-1}u = \gamma_k u = 0\}$. For $f_{k-1} \in S(\Gamma_{k-1})$ and $f_k \in S(\Gamma_k)$ we consider the following boundary value problem: find $u \in S(\Omega_k)$ such that

$$\begin{cases} a(u,v)_{\Omega_k} = 0, & \forall v \in S_0(\Omega_k), \\ \gamma_{k-1}u = f_{k-1}, \quad \gamma_k u = f_k. \end{cases} \quad (2.10.2)$$

It is assumed that (2.10.2) admits a unique solution. Let f_{k-1} and f_k be arbitrary, then the set of solutions to the problem (2.10.2) is a linear space, which is a subspace of $S(\Omega_k)$, denoted by Y_k.

By similarity of the mesh, there is a natural isomorphism among $S(\Gamma_0), S(\Gamma_1), \cdots,$ $S(\Gamma_k), \cdots$. Taking an arbitrary $f = (f^{(0)}, f^{(1)}, \ldots, f^{(m-1)}) \in S(\Gamma_k)$, where $f^{(0)}$, $f^{(1)}, \ldots, f^{(m-1)}$ are the traces from order zero to $(m-1)$ respectively, we establish a transform of independent variables, $x \to \xi^{-k}x$, then $(f^{(0)}, \xi^k f^{(1)}, \ldots, \xi^{(m-1)k}f^{(m-1)}) \in S(\Gamma_0)$ associated with the new independent variables. This isomorphism is denoted by $Z_k : S(\Gamma_k) \to S(\Gamma_0)$. Besides, $S(\Gamma_0)$ is finite dimensional. Let its dimension be n, then each $f \in S(\Gamma_0)$ corresponds to a n-dimensional column vector y. According to this relation each element in $S(\Gamma_k)$ corresponds to a n-dimensional column vector. Let $u \in S(\Omega_k)$, then the relationship of $S(\Omega_k) \to S(\Gamma_{k-1}) \to S(\Gamma_0) \to \mathbb{R}^n$ is denoted by $y = B_{k-1}u$. In like manner we can define $B_k u$.

The infinite element spaces are defined as

$$S(\Omega) = \left\{ u \in (H^m(\Omega))^N; u|_{\Omega_k} \in S(\Omega_k) \right\},$$

$$S_0(\Omega) = \{u \in S(\Omega); B_0 u = 0\}.$$

We will always assume that $(P_{m-1}(\Omega))^N \subset S(\Omega)$. Consider the following infinite element problem: for a given $y_0 \in \mathbb{R}^n$ find $u \in S(\Omega)$ such that $B_0 u = y_0$ and

$$a(u,v) = 0, \quad \forall v \in S_0(\Omega). \quad (2.10.3)$$

To establish the theory parallel with that in the sections from §2.1 to §2.3, we discuss the following typical problem first: Consider the equation

$$\triangle^m u = 0.$$

The corresponding bilinear functional is

$$a(u,v) = \sum_{|\alpha|=m} \int_\Omega a_\alpha \partial^\alpha u \partial^\alpha v \, dx,$$

where a_α are the coefficients of the expression of d terms, $(\frac{\partial^2}{\partial x_1^2} + \cdots + \frac{\partial^2}{\partial x_d^2})^m$. We have an inequality of Friedrichs type,

$$a(u,u) \geq C^{-1}\|u\|_{m,\Omega}^2, \quad \forall u \in S_0(\Omega), \quad (2.10.4)$$

where $C > 0$. Therefore the corresponding problem (2.10.3) admits a unique solution. Let u be a solution to the problem (2.10.2) with $k = 1$ and boundary condition $B_0 u = y_0$, and $B_1 u = y_1$, where $y_0, y_1 \in \mathbb{R}^n$ are arbitrary, then $u \in Y_1$. Y_1 is a $2n$-dimensional subspace, and the correspondence between this subspace and the set of (y_0, y_1) is one to one. Thus we have the expression,

$$a(u,u)_{\Omega_1} = (y_0^T \ y_1^T) \begin{pmatrix} K_0 & -A^T \\ -A & K_0' \end{pmatrix} \begin{pmatrix} y_0 \\ y_1 \end{pmatrix}.$$

If u is the solution to the problem (2.10.3), then there is a real matrix X, such that $y_k = X y_{k-1}, k = 1, 2, \cdots$. Denote by Q_l the set of all homogeneous complex polynomials of degree l.

Lemma 2.10.1. $\lambda = 1, \xi, \ldots, \xi^{m-1}$ are eigenvalues of the transfer matrix X, and the associated eigenspaces are $B_0 Q_0, B_0 Q_1, \ldots, B_0 Q_{m-1}$. All other eigenvalues λ satisfy $|\lambda| < \xi^{m-\frac{d}{2}}$. The elementary divisors associated with these eigenvalues satisfying $|\lambda| \geq \xi^{m-\frac{d}{2}}$ are linear.

Proof. Let λ be an eigenvalue, g be the associated eigenvector, and u be the solution to (2.10.3) with the boundary condition $B_0 u = g$. The transform of independent variables $x \to \xi^{-k+1} x$ maps Ω_k to Ω_1, and $u|_{\Omega_k}$ corresponds to a function defined on Ω_1, $u^{(k)} \in Y_1$. Because $y_{k-1} = \lambda^{k-1} y_0$ and $y_k = \lambda^{k-1} y_1$, we have $u^{(k)} = \lambda^{k-1} u^{(1)}$, thus

$$|u|_m^2 = \sum_{k=1}^{\infty} |u|_{m,\Omega_k}^2$$

$$= \sum_{k=1}^{\infty} |\lambda|^{2(k-1)} \xi^{(d-2m)(k-1)} |u^{(1)}|_{m,\Omega_1}^2. \quad (2.10.5)$$

Since $u \in (H^m(\Omega))^N$, the above series converges. Consequently we have either $|u^{(1)}|_{m,\Omega_1}^2 = 0$ or $|\lambda|^2 \xi^{d-2m} < 1$. For the former case, $u^{(1)} \in (P_{m-1}(\Omega_1))^N$ and it is easy to see that u is a homogeneous polynomial, for the later case $|\lambda| < \xi^{m-\frac{d}{2}}$.

If $|\lambda| \geq \xi^{m-\frac{d}{2}}$, according to the above argument, $\lambda \in \{1, \xi, \ldots, \xi^{m-1}\}$. Let us prove that the associated elementary divisor is linear. If it were not the case, then there would be a $g_1 \in \mathbb{R}^n$, such that

$$X g_1 = g + \lambda g_1.$$

By induction we would have

$$X^k g_1 = k \lambda^{k-1} g + \lambda^k g_1, \qquad k = 1, 2, \cdots.$$

Taking $B_0 u = g_1$ as the boundary condition, we would have

$$y_k = k \lambda^{k-1} g + \lambda^k y_0,$$

2.10 GENERAL ELLIPTIC SYSTEMS

hence
$$y_{k+1} - \lambda y_k = \lambda^k g. \tag{2.10.6}$$

$u^{(k)}$ could be constructed in a manner similar to the previous paragraph, then by (2.10.6) it could be proved that $u^{(k)} - \lambda^{k-1} u^{(1)} \in (P_{m-1}(\Omega_1))^N$, hence (2.10.5) would hold too. But now $|u^{(1)}|^2_{m,\Omega_1} \neq 0$ and $|\lambda| \geq \xi^{m-\frac{d}{2}}$, so the series (2.10.5) would diverge. This is a contradiction. □

We define two subspaces of $S(\Omega)$:

$$T_k(\Omega) = \left\{ u \in S(\Omega); u\big|_{\Omega_j} = 0, j \neq k, j \neq k+1 \right\}, \quad k = 1, 2, \cdots.$$

If $m - \frac{d}{2} \leq 0$, let $W(\Omega) = \{0\}$. If $m - \frac{d}{2} > 0$, we take q to be the greatest one among the integers less than $m - \frac{d}{2}$, and let

$$W(\Omega) = \left\{ u \in S(\Omega); B_0 u = 0, u\big|_{\xi\Omega} \in (P_q(\xi\Omega))^N \right\}.$$

Now we are in a position to prove the denseness theorem. Let $U_i(\Omega) = W(\Omega) \oplus T_1(\Omega) \oplus \cdots \oplus T_i(\Omega)$, $\tilde{S}_0(\Omega) = \bigcup_{i=1}^{\infty} U_i(\Omega)$.

Theorem 2.10.1. $\tilde{S}_0(\Omega)$ is dense in $S_0(\Omega)$.

Proof. Let $\Omega^* = \Omega \setminus \bar{\Gamma}_0 \setminus \{0\}$. For any $u \in S_0(\Omega)$ and any $\varepsilon > 0$, there is a $w \in W(\Omega)$ and $u_1 \in C_0^\infty(\Omega^*)$, such that $\|u - u_1 - w\|_m < \varepsilon$ [50]. We consider the following problem: find $u_2 \in S_0(\Omega)$, such that

$$a(u_2, v) = a(u_1, v), \quad \forall v \in S_0(\Omega). \tag{2.10.7}$$

(2.10.4) implies that (2.10.7) admits a unique solution. Arguing as the proof of Theorem 2.3.1, we obtain
$$\|u_2 - u_1\|_m < C\varepsilon.$$

There is a natural number k_0 such that $u_1 \equiv 0$ for $k \geq k_0$. We decompose u_2 as $u_2 = u_3 + u_4$, where u_3 corresponds to the eigenvalues $\lambda = 1, \xi, \ldots, \xi^{m-1}$ of the transfer matrix X, and u_4 corresponds to all other eigenvalues satisfying $|\lambda| < \xi^{m-\frac{d}{2}}$. Then $u_3 \in (P_{m-1}(\Omega))^N$, hence $|u_3|_m = 0$. We construct truncated functions $u_{4,k} \in S(\Omega)$ for $k \geq k_0$, such that $u_{4,k}\big|_{\Omega_{k+1}} \in Y_{k+1}$ and

$$u_{4,k} = \begin{cases} u_4, & x \in \Omega \setminus \xi^k \Omega, \\ 0, & x \in \xi^{k+1}\Omega. \end{cases}$$

Under the transform of independent variables $x \to \xi^{-k}x$, $u_{4,k}\big|_{\Omega_{k+1}}$ corresponds to a function defined on Ω_1, denoted by $u^{(k)}$. Thus

$$|u_{4,k}|_{m,\Omega_{k+1}} = \xi^{(\frac{d}{2}-m)k} |u^{(k)}|_{m,\Omega_1}.$$

But $|B_k u_4| < C\xi^{(m-\frac{d}{2})k}$, therefore
$$|u^{(k)}|_{m,\Omega_1} < C\xi^{(m-\frac{d}{2})k}.$$

Accordingly $|u_{4,k}|_{m,\Omega_{k+1}}$ are uniformly bounded with respect to k.

Arguing as the proof of Theorem 2.3.1 we can find $u_5 \in T_1(\Omega) \oplus \cdots \oplus T_i(\Omega)$, such that $|u_1 - u_5|_m < C\varepsilon$. But $u_1 - u_5 \in (H_0^m(\Omega))^N$, by the inequality of Friedrichs type we obtain
$$\|u_1 - u_5\|_m < C\varepsilon,$$
then we have
$$\|u - u_5 - w\|_m < C\varepsilon.$$

□

We return now to consider the general bilinear functional (2.10.1). Applying Theorem 2.10.1 we conclude that the problem (2.10.3) is equivalent to: find $u \in S(\Omega)$, such that $B_0 u = y_0$, and
$$a(u,v) = 0, \qquad \forall v \in W(\Omega), \forall v \in T_k(\Omega), k = 1, 2, \cdots. \tag{2.10.8}$$

We assume that the problem (2.10.3) or (2.10.8) admits a unique solution, then there is a real matrix X, such that $y_k = X y_{k-1}$, $k = 1, 2, \cdots$. And Lemma 2.10.1 is also valid for this general case. Let
$$V(\Omega) = \left\{ u \in S(\Omega); u\big|_{\Omega_k} \in Y_k, k = 1, 2, \ldots \right\},$$
$$V_0(\Omega) = \{ u \in V(\Omega); B_0 u = 0 \},$$
$$Y_k(\Omega) = T_k(\Omega) \bigcap V(\Omega), \quad W_0(\Omega) = W(\Omega) \bigcap V(\Omega).$$

The following lemma gives another equivalent formulation of the problem (2.10.3).

Lemma 2.10.2. *The problem (2.10.3) is equivalent to: find $u \in V(\Omega)$, such that $B_0 u = y_0$, and*
$$a(u,v) = 0, \qquad \forall v \in W_0(\Omega), \forall v \in Y_k(\Omega), k = 1, 2, \cdots. \tag{2.10.9}$$

Proof. If u is a solution to the problem (2.10.3), then $u\big|_{\Omega_k} \in Y_k$, hence $u \in V(\Omega)$. Besides $W_0(\Omega)$ and $Y_k(\Omega)$ are subspaces of $S_0(\Omega)$, therefore u is a solution to (2.10.9).

Conversely, if u is a solution to (2.10.9), let us prove it is also a solution to (2.10.8). Taking an arbitrary $v \in T_k(\Omega)$, we make a decomposition on Ω_k, $v\big|_{\Omega_k} = v_1 + v_2$, such that $v_1 \in Y_k$ and $v_2 \in S_0(\Omega_k)$. The same decomposition is made on Ω_{k+1}. By (2.10.9) we have
$$a(u, v_1)_{\Omega_k} + a(u, v_1)_{\Omega_{k+1}} = 0.$$

Besides $u|_{\Omega_k} \in Y_k$, therefore

$$a(u, v_2)_{\Omega_k} = 0, \quad a(u, v_2)_{\Omega_{k+1}} = 0.$$

By addition we get $a(u,v)_\Omega = 0$. In the same way, taking an arbitrary $v \in W(\Omega)$, we make a decomposition on Ω_1, $v|_{\Omega_1} = v_1 + v_2$, such that $v_1 \in Y_1$, $v_2 \in S_0(\Omega_1)$. By (2.10.9) we have

$$a(u,v)_{\xi\Omega} + a(u,v_1)_{\Omega_1} = 0.$$

Besides $u|_{\Omega_1} \in Y_1$, therefore

$$a(u, v_2)_{\Omega_1} = 0.$$

By addition we get $a(u,v)_\Omega = 0$ too. □

We take any $u, v \in Y_k$ and let $y_{k-1} = B_{k-1}u$, $y_k = B_k u$, $z_{k-1} = B_{k-1}v$, $z_k = B_k v$, then $a(u,v)_{\Omega_k}$ can be expressed in terms of $y_{k-1}, y_k, z_{k-1}, z_k$. Let

$$a(u,v)_{\Omega_k} = (z_{k-1}^T \quad z_k^T) \begin{pmatrix} K_0 & -B \\ -A & K_0' \end{pmatrix} \begin{pmatrix} y_{k-1} \\ y_k \end{pmatrix},$$

then by the system of equations (2.10.9) and the arbitrariness of v we obtain

$$g^T(-Ay_0 + K_0' y_1) = 0,$$

$$-Ay_0 + (K_0' + \xi^{d-2m} K_0) y_1 - \xi^{d-2m} B y_2 = 0,$$

$$\dots\dots\dots\dots$$

$$-Ay_{k-1} + (K_0' + \xi^{d-2m} K_0) y_k - \xi^{d-2m} B y_{k+1} = 0,$$

$$\dots\dots\dots\dots$$

where $g \in B_1 W(\Omega)$. Consequently we obtain the equations satisfied by the transfer matrix X,

$$g^T(-A + K_0' X) = 0, \quad \forall g \in B_1 W(\Omega), \qquad (2.10.10)$$

$$\xi^{d-2m} B X^2 - (K_0' + \xi^{d-2m} K_0) X + A = 0. \qquad (2.10.11)$$

Being analogous to Theorem 2.3.3, we have

Theorem 2.10.2. *The necessary and sufficient conditions to determine the transfer matrix X are*

(a) *if λ is an eigenvalue then either $\lambda = \xi^l$, l is a non-negative integer, $l \leq m - \frac{d}{2}$, and the associated eigenspace is $B_0 Q_l$, the associated elementary divisors are linear, or $|\lambda| < \xi^{m-\frac{d}{2}}$;*

(b) *X satisfies the equation (2.10.10);*

(c) *X satisfies the equation (2.10.11).*

Proof. On account of Lemma 2.10.1 and Lemma 2.10.2 the necessity is obvious, let us prove the sufficiency.

It is assumed that X satisfies (a),(b),(c). We take an arbitrary $y_0 \in \mathbb{R}^n$ and set $y_k = X^k y_0$. The problem (2.10.2) is solved on all domains Ω_k. By Lemma 2.10.2, it will suffice to prove the obtained solution $u \in (H^m(\Omega))^N$.

We make a decomposition $y_0 = y_0^{(1)} + y_0^{(2)}$, where $y_0^{(1)}$ belongs to the invariant subspace associated with the eigenvalues $\lambda = \xi^l$ of the transfer matrix X, and $y_0^{(2)}$ belongs to the invariant subspace associated with $|\lambda| < \xi^{m-\frac{d}{2}}$. It leads to a decomposition $u = u^{(1)} + u^{(2)}$. By the uniqueness of the solution to the problem (2.10.2), $u^{(1)} \in (P_{m-1}(\Omega))^N \subset (H^m(\Omega))^N$. It will suffice to prove $u^{(2)} \in (H^m(\Omega))^N$.

Under a transform of independent variables $x \to \xi^{-k+1} x, \Omega_k \to \Omega_1$. If $u|_{\Omega_k} \to u^{(k)} \in S(\Omega_1)$ under this transform, then

$$|u|^2_{m,\Omega_k} = \xi^{(d-2m)(k-1)} |u^{(k)}|^2_{m,\Omega_1},$$

therefore

$$|u|^2_{m,\Omega} = \sum_{k=1}^{\infty} |u|^2_{m,\Omega_k} = \sum_{k=1}^{\infty} \xi^{(d-2m)(k-1)} |u^{(k)}|^2_{m,\Omega_1}.$$

Since the space $S(\Omega_1)$ is finite dimensional and the solution to the problem (2.10.2) uniquely exists, we have

$$|u^{(k)}|^2_{m,\Omega_1} \leq C(|y_k|^2 + |y_{k-1}|^2).$$

Then the inequality (2.3.8) yields $|u|^2_{m,\Omega} < +\infty$. □

If $a(u,v)$ is a symmetric and semi-positive definite bilinear functional, the iterative method of the first type is effective. According to the recurrence formulas,

$$K_{l+1} = K_l - A_l^T (K_l' + \xi^{(d-2m)k} K_l)^{-1} A_l,$$
$$K_{l+1}' = K_l' - \xi^{2(d-2m)k} A_l (K_l' + \xi^{(d-2m)k} K_l)^{-1} A_l^T,$$
$$A_{l+1} = \xi^{(d-2m)k} A_l (K_l' + \xi^{(d-2m)k} K_l)^{-1} A_l,$$

we can get the combined stiffness matrix on $\Omega \setminus \xi^{2^l}\Omega$,

$$\begin{pmatrix} K_l & -A_l^T \\ -A_l & K_l' \end{pmatrix}$$

. Then we define a subspace

$$D_k(\Omega) = \left\{ u \in S(\Omega); u|_{\xi^{2^l}\Omega} \in (P_{m-1}(\xi^{2^l}\Omega))^N \right\},$$

and the approximate combined stiffness matrix $K_z^{(l)}$ is thus obtained. The argument is the same as the previous sections, we would not repeat it here.

2.11 Exterior Stokes problems (II)

This section is devoted to give a rigorous proof of the algorithm in §1.10. Let $x = (x_1, x_2)$ be a point in $\mathbb{R}^2$, Ω be the domain in §1.10. Let $u = (u_1, u_2)$ denote the velocity and p the pressure, then the considered governing equations are

$$\mu \left(-\frac{1}{x_1} \nabla(x_1 \nabla u_1) + \frac{1}{x_1^2} u_1 \right) + \frac{\partial p}{\partial x_1} = 0,$$

$$-\frac{\mu}{x_1} \nabla(x_1 \nabla u_2) + \frac{\partial p}{\partial x_2} = 0,$$

$$\frac{\partial}{\partial x_1}(x_1 u_1) + \frac{\partial}{\partial x_2}(x_1 u_2) = 0,$$

where the constant $\mu > 0$ is viscosity. We define bilinear functionals

$$a(u,v) = \mu \int x_1 \left(\nabla u_1 \cdot \nabla v_1 + \nabla u_2 \cdot \nabla v_2 + \frac{u_1 v_1}{x_1^2} \right) dx, \qquad (2.11.1)$$

$$b(u,p) = \int p \left(\frac{\partial}{\partial x_1}(x_1 u_1) + \frac{\partial}{\partial x_2}(x_1 u_2) \right) dx. \qquad (2.11.2)$$

Introduce norms

$$\|p\|_0^2 = \int_\Omega x_1 p^2(x) \, dx,$$

$$\|u_2\|_1^2 = \int_\Omega x_1 \left(|\nabla u_2(x)|^2 + \frac{u_2^2(x)}{|x|^2} \right) dx,$$

$$\|u_1\|_{1,\star}^2 = \int_\Omega x_1 \left(|\nabla u_1(x)|^2 + \frac{u_1^2(x)}{x_1^2} \right) dx,$$

and corresponding Hilbert spaces $Z^0(\Omega), Z^1(\Omega), Z^{1,\star}(\Omega)$.

After partitioning Ω as Fig. 17, we introduce some infinite element spaces,

$$S(\Omega) = \left\{ p \in Z^0(\Omega); p\big|_{e_i} \in P_0(e_i), i = 1, 2, \ldots \right\},$$

$$U(\Omega) = \Big\{ u = (u_1, u_2); u_1 \in Z^{1,\star}(\Omega), u_1\big|_{e_i} \in P_2(e_i), i = 1, 2, \ldots,$$

$$u_2 \in Z^1(\Omega), u_2\big|_{e_i} \in P_2(e_i), i = 1, 2, \ldots \Big\},$$

$$U_0(\Omega) = \{ u \in U(\Omega); u\big|_{\Gamma_0} = 0 \},$$

$$U(\Gamma_0) = \{ u\big|_{\Gamma_0}; u \in U(\Omega) \}.$$

Then we can formulize the infinite element scheme as follows: find $u \in U(\Omega)$, $p \in S(\Omega)$, such that $u|_{\Gamma_0} = f \in U(\Gamma_0)$, and u, p satisfy the equations

$$\begin{cases} a(u,v) - b(v,p) = 0, & \forall v \in U_0(\Omega), \\ b(u,q) = 0, & \forall q \in S(\Omega). \end{cases} \quad (2.11.3)$$

We costruct some subspaces of $U(\Omega)$,

$$V(\Omega) = \{u \in U(\Omega); b(u,p) = 0, \forall p \in S(\Omega)\},$$

$$V_0(\Omega) = \{u \in V(\Omega); u|_{\Gamma_0} = 0\}.$$

Then the solution u to the problem (2.11.3) satisfies: $u|_{\Gamma_0} = f, u \in V(\Omega)$, and

$$a(u,v) = 0, \quad \forall v \in V_0(\Omega). \quad (2.11.4)$$

By §2.10 there is a combined stiffness matrix K_z, which is symmetric and semi-positive definite, for the problem (2.11.4). We will prove that it is positive definite later on.

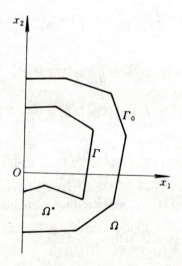

Fig. 27

Now we consider a more general exterior domain Ω_0. If Ω_0 is decomposed to Ω^* and Ω by a broken line Γ_0, where Ω^* is a bounded domain (Fig. 27). Conventional partition is taken on Ω^*. Analogically we define spaces $S(\Omega_0), U(\Omega_0), U_0(\Omega_0), U(\Gamma)$, and $V(\Omega_0), V_0(\Omega_0)$. Taking $f \in U(\Gamma)$, we can put forward the following problem: find $u \in V(\Omega_0)$, such that $u|_\Gamma = f$, and

$$a(u,v) = 0, \quad \forall v \in V_0(\Omega_0). \quad (2.11.5)$$

Lemma 2.11.1. $a(\cdot,\cdot)$ is symmetric and positive definite on $U_0(\Omega_0)$.

Proof. By (2.11.1)
$$a(u,u)_{\Omega_0} = \mu \int_{\Omega_0} x_1 \left(|\nabla u_1|^2 + |\nabla u_2|^2 + \frac{u_1^2}{x_1^2}\right) dx.$$

We return to three dimension by defining
$$v(x_1, x_2, x_3) = u_2\left(\sqrt{x_1^2 + x_3^2}, x_2\right),$$

which is defined on a three dimensional domain $\tilde{\Omega}_0$. We have
$$2\pi \int_{\Omega_0} x_1 |\nabla u_2|^2 \, dx = \int_{\tilde{\Omega}_0} |\nabla v|^2 \, dx.$$

According to the definition of the space $Z^1(\Omega)$,
$$\int_{\tilde{\Omega}_0} \left(|\nabla v|^2 + \frac{v^2(x)}{|x|^2}\right) dx = 2\pi \|u_2\|_1^2 < +\infty,$$

so Lemma 2.1.1 implies that $v \in H_0^{1,*}(\tilde{\Omega}_0)$. Applying the inequality (2.1.1) we get
$$\|v\|_{1,*,\tilde{\Omega}_0}^2 \leq C \int_{\tilde{\Omega}_0} |\nabla v|^2 \, dx,$$

hence
$$\|u_2\|_1^2 \leq C \int_{\Omega_0} x_1 |\nabla u_2|^2 \, dx.$$

Consequently
$$a(u,u)_{\Omega_0} \geq C^{-1}(\|u_1\|_{1,*}^2 + \|u_2\|_1^2).$$

$\square$

Lemma 2.11.1 implies that the problem (2.11.5) admits a unique solution.
Let $y_0 = B_0 u, z_0 = B_0 v$, where u is the solution to the problem (2.11.5), then
$$a(u,v)_{\Omega_0} = a(u,v)_{\Omega^*} + z_0^T K_z y_0. \tag{2.11.6}$$

Analogically we may introduce finite element spaces $S(\Omega^*)$, $U(\Omega^*)$, and
$$U_0(\Omega^*) = \{u \in U(\Omega^*); u|_\Gamma = 0\}.$$

The solution u to the problem (2.11.5) is restricted by
$$b(u,p)_{\Omega^*} = 0, \qquad \forall p \in S(\Omega^*)$$

on the domain Ω^*. Regarding p as a Lagrangian multiplier, we can get a new formulation of (2.11.5): find $u \in U(\Omega^*), p \in S(\Omega^*)$, such that $u|_\Gamma = f$ and
$$a(u,v)_{\Omega^*} + z_0^T K_z y_0 - b(v,p)_{\Omega^*} = 0, \quad \forall v \in U_0(\Omega^*), \tag{2.11.7}$$
$$b(u,q)_{\Omega^*} = 0, \quad \forall q \in S(\Omega^*). \tag{2.11.8}$$

Lemma 2.11.2. *There is a unique solution to the problem* (2.11.7),(2.11.8).

Proof. We prove the uniqueness first. Let (u,p) be the solution to the corresponding homogeneous problem. Take $q = p$ in (2.11.8) and $v = u$ in (2.11.7). Upon summing these two expressions we obtain

$$a(u,u)_{\Omega^*} + y_0^T K_z y_0 = 0.$$

Since $a(\cdot,\cdot)_{\Omega^*}$ and K_z are positive definite, $u = 0$. Then we take such a v in (2.11.7) that it is nonvanishing at the middle point of one interior side and vanishes at all other nodes. From (2.11.2) and Green's formula it is deduced that p is a same constant on both two adjacent elements of this side. Thus p is a constant on Ω^*. We take $v \in U_0(\Omega^*)$ again, such that

$$\int_{\Omega^*} \left(\frac{\partial}{\partial x_1}(x_1 v_1) + \frac{\partial}{\partial x_2}(x_1 v_2) \right) dx \neq 0,$$

then (2.11.2) and (2.11.7) imply $p = 0$.

(2.11.7),(2.11.8) is a system of algebraic equations with finite order, and the matrix of coefficients is square, therefore the uniqueness of solution implies existence for any right hand side. □

Theorem 2.11.1. *The problem* (2.11.5) *is equivalent to* (2.11.7),(2.11.8).

Proof. The problem (2.11.5) can be rewritten as

$$a(u,u)_{\Omega_0} = \min_{\substack{v \in V(\Omega_0) \\ v|_\Gamma = f}} a(v,v).$$

For any $v \in V(\Omega^*)$ we extend v to Ω such that it is the solution to the problem (2.11.4) and the continuity of v on Γ_0 is maintained. This extension is unique because so is the solution to (2.11.4). The set of functions obtained in this way is denoted by P, then the solutions to the the problem (2.11.5), $u \in P$. Therefore

$$a(u,u)_{\Omega_0} = \min_{\substack{v \in P \\ v|_\Gamma = f}} a(v,v)_{\Omega_0}.$$

Noting (2.11.6) and introducing a Lagrangian multiplier p, we get (2.11.7),(2.11.8). □

Now we verify the algorithm in §1.10. First of all, the solutions expressed by (1.10.5),(1.10.6) uniquely exist. The proof of this assertion is the same as that of Lemma 2.11.2, we would not repeat it here. But we would emphasize that this solution is physically meaningful only if y_0 and y_1 satisfy (1.10.8). Otherwise an incompressible flow is impossible, and the obtained y^*, q only exist as some intermediate variables of the procedure. Besides, we have set the pressure to be zero on one assigned element. If y_0 and y_k satisfy (1.10.8), then denote by p_0 the value of p

on the assigned element for any $p \in S(\Omega \setminus \overline{\xi^k \Omega})$, $p - p_0$ is the desired one. Because p_0 is a constant, we have

$$b(u, p - p_0)_{\Omega \setminus \overline{\xi^k \Omega}} = 0, \quad b(u, p_0)_{\Omega \setminus \overline{\xi^k \Omega}} = 0,$$

which leads to
$$b(u, p)_{\Omega \setminus \overline{\xi^k \Omega}} = 0, \quad \forall p \in S(\Omega \setminus \overline{\xi^k \Omega}). \tag{2.11.9}$$

Therefore (2.11.9) actually holds for any piecewise constant function p. If y_0 and y_k do not satisfy (1.10.8), then the above argument is untenable. It is why we should introduce restrictions (1.10.8),(1.10.14) and (1.10.16) for the iterative scheme.

We are now in a position to prove the convergence of the iterative method of the first type in §1.10.

Lemma 2.11.3. *All eigenvalues λ of the transfer matrix X satisfy $|\lambda| < \xi^{-\frac{1}{2}}$.*

Proof. Let λ be an eigenvalue, g be the associated eigenvector, we set $y_0 = g$. Denote by u the solution to the problem (2.11.4), then

$$y_k = X^k y_0 = \lambda^k g, \quad k = 1, 2, \cdots.$$

By similarity we have

$$a(u, u)_\Omega = \sum_{k=1}^\infty a(u, u)_{\Omega_k} = \sum_{k=1}^\infty \xi^{k-1} \lambda^{2(k-1)} a(u, u)_{\Omega_1}.$$

Since $a(u, u)_\Omega$ is bounded, we get $|\lambda| < \xi^{-\frac{1}{2}}$. □

We take and fix a $g \in \mathbb{R}^{2N}$ and take an arbitrary natural number k, then we construct a subspace of $V(\Omega)$,

$$D_k(\Omega) = \{u \in V(\Omega); B_k u = \alpha g, \alpha \in \mathbb{R}\}.$$

Taking $f \in U(\Gamma_0)$, we construct $\bar{u}_k \in D_k(\Omega)$, such that $\bar{u}_k\big|_{\Gamma_0} = f$ and

$$a(\bar{u}_k, \bar{u}_k) = \min_{\substack{v \in D_k(\Omega) \\ v|_{\Gamma_0} = f}} a(v, v). \tag{2.11.10}$$

The problem (2.11.10) admits a unique solution. Denote by K_z the combined stiffness matrix on Ω, then

$$a(\bar{u}_k, \bar{u}_k) = a(\bar{u}_k, \bar{u}_k)_{\Omega \setminus \overline{\xi^k \Omega}} + \xi^k (\alpha g)^T K_z (\alpha g). \tag{2.11.11}$$

There is a symmetric matrix $\bar{K}_z^{(k)}$, such that

$$a(\bar{u}_k, \bar{u}_k)_{\Omega \setminus \overline{\xi^k \Omega}} = y_0^T \bar{K}_z^{(k)} y_0. \tag{2.11.12}$$

Theorem 2.11.2.
$$\lim_{k \to \infty} \bar{K}_z^{(k)} = K_z. \tag{2.11.13}$$

Proof. We define a space analogous to $D_k(\Omega)$,

$$D_k(\xi^{k-1}\Omega) = \{u \in V(\xi^{k-1}\Omega); B_k u = \alpha g, \alpha \in \mathbb{R}\}.$$

Let u be the solution to the problem (2.11.4), $y_m = B_m u, m = 0, 1, \cdots$. We construct $w^{(k)} \in V(\Omega)$, such that $w^{(k)}|_{\Omega \setminus \overline{\xi^{k-1}\Omega}} = u|_{\Omega \setminus \overline{\xi^{k-1}\Omega}}$, $w^{(k)}|_{\xi^{k-1}\Omega} \in D_k(\xi^{k-1}\Omega)$, and

$$a(w^{(k)}, w^{(k)})_{\xi^{k-1}\Omega} = \min_{\substack{B_{k-1}v = y_{k-1} \\ v \in D_k(\xi^{k-1}\Omega)}} a(v, v)_{\xi^{k-1}\Omega}.$$

By (1.10.14), α is uniquely determined by y_0. We take y_0 and αg as the boundary values, then $\bar{u}_k$ makes the quadratic functional $a(\cdot, \cdot)$ reach its minimum value on $\Omega \setminus \overline{\xi^k \Omega}$. Hence

$$\begin{aligned}
a(\bar{u}_k, \bar{u}_k)_{\Omega \setminus \overline{\xi^k \Omega}} &\leq a(w^{(k)}, w^{(k)})_{\Omega \setminus \overline{\xi^k \Omega}} \\
&= a(u, u)_{\Omega \setminus \overline{\xi^{k-1}\Omega}} + a(w^{(k)}, w^{(k)})_{\Omega_k} \\
&= a(u, u)_\Omega - a(u, u)_{\xi^{k-1}\Omega} + a(w^{(k)}, w^{(k)})_{\Omega_k}.
\end{aligned} \tag{2.11.14}$$

By Lemma 2.11.3 we have the inequalities

$$|y_{k-1}| < C\eta^{k-1}, \quad \eta < \xi^{-\frac{1}{2}},$$

therefore

$$\begin{aligned}
a(u, u)_{\xi^{k-1}\Omega} &= \xi^{k-1} y_{k-1}^T K_z y_{k-1} \\
&\leq C\xi^{k-1}\eta^{2(k-1)} \to 0 \quad (k \to \infty).
\end{aligned}$$

Besides

$$a(w^{(k)}, w^{(k)})_{\Omega_k} = \xi^{k-1} \begin{pmatrix} y_{k-1}^T & \alpha g^T \end{pmatrix} \begin{pmatrix} K_0 & -A^T \\ -A & K_0' \end{pmatrix} \begin{pmatrix} y_{k-1} \\ \alpha g \end{pmatrix},$$

and from (1.10.14) we get

$$|\alpha| \leq C\xi^{-2k}, \tag{2.11.15}$$

which yields

$$a(w^{(k)}, w^{(k)})_{\Omega_k} \to 0 \quad (k \to \infty).$$

$D_k(\Omega)$ is a subspace of $V(\Omega)$, therefore

$$a(\bar{u}_k, \bar{u}_k) \geq a(u, u). \tag{2.11.16}$$

(2.11.15) leads to
$$\xi^k(\alpha g)^T K_z(\alpha g) \to 0 \quad (k \to \infty).$$

Combining (2.11.11),(2.11.14),(2.11.16) and the above limit relations we obtain
$$\lim_{k \to \infty} a(\bar{u}_k, \bar{u}_k)_{\Omega \setminus \overline{\xi^k \Omega}} = a(u, u),$$

the matrix form of which is just (2.11.13). □

We turn now to prove the convergence of the iterative method of the second type. Given a symmetric and positive definite matrix $K_z^{(0)}$, we construct a quadratic functional on $V(\Omega)$ for $k \geq 1$ as follows:
$$a_k(u, u) = a(u, u)_{\Omega \setminus \overline{\xi^k \Omega}} + \xi^k y_k^T K_z^{(0)} y_k, \qquad (2.11.17)$$

where $y_k = B_k u$. For $f \in U(\Gamma_0)$, u_k is the solution to the following problem, which is unique on the domain $\Omega \setminus \overline{\xi^k \Omega}$, $u_k|_{\Gamma_0} = f$, and
$$a_k(u_k, u_k) = \min_{\substack{v \in V(\Omega) \\ v|_{\Gamma_0} = f}} a_k(v, v). \qquad (2.11.18)$$

There is a symmetric matrix $K_z^{(k)}$, such that
$$a_k(u_k, u_k) = y_0^T K_z^{(k)} y_0.$$

Theorem 2.11.3.
$$\lim_{k \to \infty} K_z^{(k)} = K_z. \qquad (2.11.19)$$

Proof. According to the definition of the space $D_k(\Omega)$ and (2.11.7) we have
$$a_k(\bar{u}_k, \bar{u}_k) = a(\bar{u}_k, \bar{u}_k)_{\Omega \setminus \overline{\xi^k \Omega}} + \xi^k(\alpha g)^T K_z^{(0)}(\alpha g).$$

(2.11.12) and (2.11.13) imply that
$$a(\bar{u}_k, \bar{u}_k)_{\Omega \setminus \overline{\xi^k \Omega}} \leq C,$$

and (2.11.15) implies
$$|\xi^k(\alpha g)^T K_z^{(0)}(\alpha g)| \leq C\xi^{-3k}. \qquad (2.11.20)$$

In view of (2.11.18) we get
$$a_k(u_k, u_k) \leq a_k(\bar{u}_k, \bar{u}_k) \leq C.$$

We extend the function u_k on $\xi^k \Omega$ such that u_k vanishes at all nodes except those on Γ_k, then we obtain a function, still denoted by u_k, defined on Ω. Using (2.11.20) we have $a(u_k, u_k) \leq C$. Then following the argument in the proof of Theorem 2.9.2, we obtain (2.11.19). □

2.12 Nonhomogeneous equations and the Helmholtz equation

We consider a polygonal domain Ω shown in Fig.25 of §2.4. Let $p \in L^2(\Omega)$. Consider the equation,

$$-\Delta u = p, \qquad (2.12.1)$$

on Ω, and the boundary conditions,

$$u\big|_{\Gamma^*} = u\big|_{\Gamma_*} = 0, \qquad (2.12.2)$$

$$u\big|_{\Gamma_0} = 0. \qquad (2.12.3)$$

After making a similar and infinite partition on Ω, let

$$S_0(\Omega) = \left\{ u \in H^1(\Omega); u\big|_{e_i} \in P_1(e_i), i = 1, 2, \ldots, u\big|_{\partial\Omega} = 0 \right\},$$

then the infinite element solution u satisfies $u \in S_0(\Omega)$ and

$$a(u, v) = (p, v), \qquad \forall v \in S_0(\Omega), \qquad (2.12.4)$$

where $a(\cdot, \cdot)$ is defined by (2.1.6), the corresponding algebraic system of infinite equations is just (1.11.1),

$$-A y_{k-1} + K y_k - A^T y_{k+1} = p_k, \quad k = 1, 2, \cdots. \qquad (2.12.5)$$

Let b_i be the i-th node on Γ_k. Define a shape function $\varphi \in S_0(\Omega)$, where φ assumes value 1 at the point b_i and vanishes at all other nodes. Then the i-th component of the vector p_k is

$$p_k^{(i)} = (p, \varphi).$$

Let $D = \operatorname{supp} \varphi$, then by the Schwarz inequality

$$|p_k^{(i)}| \leq \|p\|_{0,D} \|\varphi\|_{0,D} \leq \|p\|_{0,\Omega} \|\varphi\|_{0,D}.$$

It is easy to see that the area of D is bounded by $C\xi^{2k}$, therefore

$$|p_k^{(i)}| \leq C\xi^k \|p\|_{0,\Omega}. \qquad (2.12.6)$$

2.12 NONHOMOGENEOUS AND HELMHOLTZ EQUATIONS

We consider an auxiliary algebraic system of infinite number of equations,

$$\begin{cases} -Az_0^{(m)} + Kz_1^{(m)} - A^T z_2^{(m)} = 0, \\ \quad \cdots\cdots\cdots\cdots \\ -Az_{m-2}^{(m)} + Kz_{m-1}^{(m)} - A^T z_m^{(m)} = 0, \\ -Az_{m-1}^{(m)} + Kz_m^{(m)} - A^T z_{m+1}^{(m)} = p_m, \\ -Az_m^{(m)} + Kz_{m+1}^{(m)} - A^T z_{m+2}^{(m)} = 0, \\ \quad \cdots\cdots\cdots\cdots \end{cases} \quad (2.12.7)$$

and seek a particular solution to it. In the following we omit the superscript m for writing simplicity. This particular solution might be constructed as follows:

$$z_{k+1} = X z_k, \qquad \text{as } k \geq m, \quad (2.12.8)$$

$$z_{k-1} = \tilde{X} z_k, \qquad \text{as } k \leq m, \quad (2.12.9)$$

where the transfer matrix $\tilde{X}$ corresponds to the associated exterior problem. (Since the problem considered here is distinct from §2.6, the positions of $X, \tilde{X}$ just change comparing with the transfer matrices in §2.6.) Thus all homogeneous equations in (2.12.7) are satisfied, and we simply need only to seek z_m. Let $u^{(m)}$ be the interpolating function of $\{z_k\}$, then by the properties of the transfer matrix X we know $u^{(m)} \in S(\Omega)$. Substituting (2.12.8),(2.12.9) into (2.12.7) we get

$$(K - A\tilde{X} - A^T X) z_m = p_m. \quad (2.12.10)$$

Lemma 2.12.1. *The system (2.12.10) admits a unique solution z_m, satisfying*

$$|z_m| \leq C |p_m|. \quad (2.12.11)$$

Proof. Letting the equations of the two adjacent sides Γ^*, Γ_* of the point O be $\vartheta = 0$ and $\vartheta = \alpha$, we consider a sector $\Omega^\infty = \{(r,\vartheta); 0 < r < \infty, 0 < \vartheta < \alpha\}$. A space $S(\Omega^\infty)$ is constructed on the analogy of $S(\Omega)$.

To prove the existence and uniqueness of the solution to (2.12.10), it will suffice to prove the corresponding homogeneous equation admits only null solution. Let z_m be a solution to the homogeneous equation. We can determine $z_k, -\infty < k < +\infty$, by (2.12.8),(2.12.9), then obtain the interpolating function $u^{(0)} \in S(\Omega^\infty)$, which satisfies

$$a(u^{(0)}, v) = 0, \qquad \forall v \in S(\Omega^\infty).$$

Taking $v = u^{(0)}$, we obtain $a(u^{(0)}, u^{(0)}) = 0$, thus $u^{(0)} = 0$, hence $z_m = 0$.

Let us prove the inequality (2.12.11). By (2.12.10) we get

$$z_m = (K - A\tilde{X} - A^T X)^{-1} p_m,$$

which leads to (2.12.11). □

Lemma 2.12.2. *The series $u_2 = \sum_{m=1}^{\infty} u^{(m)}$ converges in both spaces $C(\bar{\Omega})$ and $H^1(\Omega)$, and*

$$\|u_2\|_{0,\infty,\Omega} + \|u_2\|_{1,\Omega} \le C\|p\|_{0,\Omega}.$$

Proof. Combining (2.12.6),(2.12.11) gives

$$|z_m| \le C\xi^m \|p\|_{0,\Omega}. \tag{2.12.12}$$

From the properties of the transfer matrices X and $\tilde{X}$ we conclude that

$$\|u^{(m)}\|_{0,\infty,\Omega} \le C\xi^m \|p\|_{0,\Omega}.$$

Then the convergence and estimate in $C(\bar{\Omega})$ is obtained.

Under the similarity transformation $x \to \xi^{-m} x$, Γ_m is transferred to Γ_0. Let $\tilde{u}^{(m)}(x) = u^{(m)}(\xi^m x)$, then $\tilde{u}^{(m)}$ is an infinite element solution on both Ω and the exterior of Γ_0. By (2.12.12) we get

$$\left(\int_{\Omega^\infty} |\nabla \tilde{u}^{(m)}|^2 \, dx\right)^{\frac{1}{2}} \le C\xi^m \|p\|_{0,\Omega}.$$

The above transformation of independent variables does not interfare in the value of integration, hence

$$\left(\int_{\Omega^\infty} |\nabla u^{(m)}|^2 \, dx\right)^{\frac{1}{2}} \le C\xi^m \|p\|_{0,\Omega}.$$

By the Friedrichs inequality

$$\|u^{(m)}\|_{1,\Omega} \le C\xi^m \|p\|_{0,\Omega}.$$

By this the convergence and estimate in $H^1(\Omega)$ is obtained. □

So far we have indeed obtained a particular solution to the equation (2.12.1) and the boundary condition (2.12.2). According to the approach in §1.11, we can find the solution to the boundary value problem.

Now we change the boundary condition (2.12.2) for the Neumann condition,

$$\left.\frac{\partial u}{\partial \nu}\right|_{\Gamma^*} = \left.\frac{\partial u}{\partial \nu}\right|_{\Gamma_*} = 0, \tag{2.12.13}$$

and the corresponding space is

$$S_0(\Omega) = \left\{ u \in H^1(\Omega); u\big|_{e_i} \in P_1(e_i), i = 1, 2, \ldots, u\big|_{\Gamma_0} = 0 \right\}.$$

2.12 NONHOMOGENEOUS AND HELMHOLTZ EQUATIONS

We also have the system (2.12.5). Let

$$p_0 = -\sum_{k=1}^{\infty} p_k,$$

then we determine $\tilde{p}_k$, $k = 0, 1, \ldots$, by the following recurrence relations:

$$\tilde{p}_0 = p_0,$$
$$\tilde{p}_m = \tilde{p}_{m-1} + p_m, \qquad m = 1, 2, \cdots.$$

It is easy to see that

$$\tilde{p}_m = \sum_{k=0}^{m} p_k = -\sum_{k=m+1}^{\infty} p_k.$$

Then the inequality (2.12.6) yields

$$|\tilde{p}_m| \leq C\xi^m \|p\|_{0,\Omega}. \tag{2.12.14}$$

We consider an auxiliary algebraic system of infinite number of equations, $m = 1, 2, \ldots$,

$$-Az_0^{(m)} + Kz_1^{(m)} - A^T z_2^{(m)} = 0,$$
$$\cdots\cdots\cdots\cdots$$
$$-Az_{m-3}^{(m)} + Kz_{m-2}^{(m)} - A^T z_{m-1}^{(m)} = 0,$$
$$-Az_{m-2}^{(m)} + Kz_{m-1}^{(m)} - A^T z_m^{(m)} = \tilde{p}_{m-1}, \tag{2.12.15}$$
$$-Az_{m-1}^{(m)} + Kz_m^{(m)} - A^T z_{m+1}^{(m)} = -\tilde{p}_{m-1}, \tag{2.12.16}$$
$$-Az_m^{(m)} + Kz_{m+1}^{(m)} - A^T z_{m+2}^{(m)} = 0,$$
$$\cdots\cdots\cdots\cdots$$

We can deduce the counterpart of (2.12.10),

$$(K - A\tilde{X})z_{m-1} - A^T z_m = \tilde{p}_{m-1}, \tag{2.12.17}$$

$$-Az_{m-1} + (K - A^T X)z_m = -\tilde{p}_{m-1}. \tag{2.12.18}$$

Lemma 2.12.3. *The system of equations (2.12.17),(2.12.18) admits solutions. If the last component of z_m is set to be zero, then the solution is unique and satisfies*

$$|z_{m-1}| + |z_m| \leq C|\tilde{p}_{m-1}|. \tag{2.12.19}$$

Proof. First we seek the general solutions to the corresponding homogeneous equations of (2.12.17),(2.12.18). If z_{m-1}, z_m are a couple of solutions to the homogeneous equations, let

$$z_{k+1} = Xz_k, \qquad \text{as } k \geq m, \tag{2.12.20}$$

$$z_{k-1} = \tilde{X} z_k, \qquad \text{as } k \leq m-1, \tag{2.12.21}$$

then we obtain $u^{(0)} \in S(\Omega^\infty)$ which satisfies

$$a(u^{(0)}, v) = 0, \qquad \forall v \in S(\Omega^\infty).$$

Taking $v = u^{(0)}$, we conclude that $u^{(0)}$ is a constant, that is $z_k = \alpha g_1$. On the contrary if we take $z_{m-1} = z_m = \alpha g_1$, then Lemma 2.6.1 implies that they are the solutions to the homogeneous equations, hence they are general solutions.

Now

$$\begin{pmatrix} g_1^T & g_1^T \end{pmatrix} \begin{pmatrix} \tilde{p}_{m-1} \\ -\tilde{p}_{m-1} \end{pmatrix} = 0,$$

therefore the solutions to (2.12.17),(2.12.18) exist, Let z_{m-1}, z_m be a couple of particular solutions, then the general solutions are $z_{m-1} + \alpha g_1$, $z_m + \alpha g_1$. We take the constant α appropriately, such that the last component of z_m becomes zero, then the solutions are determined uniquely. Taking the inverse matrix we can prove the inequality (2.12.19). □

Let the interpolating function of $\{z_k^{(m)}\}$ be $u^{(m)}$, then following the proof of Lemma 2.12.2 we can prove

Lemma 2.12.4. *The series $u_2 = \sum_{m=1}^\infty u^{(m)}$ converges in both spaces $C(\bar{\Omega})$ and $H^1(\Omega)$, moreover*

$$\|u_2\|_{0,\infty,\Omega} + \|u_2\|_{1,\Omega} \leq C \|g\|_{0,\Omega}.$$

We have indeed obtained a particular solution to the equation (2.12.1) and the boundary condition (2.12.13).

Remark.

If p is a bounded function, then the estimate (2.12.6) can be improved. Now

$$|p_k^{(i)}| \leq \int_D |p|\, dx \leq C \xi^{2k} \|p\|_{0,\infty,\Omega}.$$

By Lemma 2.12.1 or Lemma 2.12.3 we get

$$|z_k^{(m)}| \leq C \xi^{2m} \|p\|_{0,\infty,\Omega}. \tag{2.12.22}$$

We turn now to the Helmholtz equation with the boundary condition (2.12.13). We construct the following subspaces of the complex Sobolev spaces,

$$S(\Omega) = \left\{ u \in H^1(\Omega); u|_{e_i} \in P_1(e_i), i = 1, 2, \ldots \right\},$$

$$S_0(\Omega) = \left\{ u \in S(\Omega); u|_{\Gamma_0} = 0 \right\},$$

$$S(\Gamma_0) = \left\{ u|_{\Gamma_0}; u \in S(\Omega) \right\}.$$

2.12 NONHOMOGENEOUS AND HELMHOLTZ EQUATIONS

Letting λ be a complex constant, we consider a sesquilinear form

$$a_\lambda(u,v) = \int_\Omega \left(\frac{\partial u}{\partial x_1} \overline{\frac{\partial v}{\partial x_1}} + \frac{\partial u}{\partial x_2} \overline{\frac{\partial v}{\partial x_2}} + \lambda u \bar{v} \right) dx$$

on $S(\Omega)$, and a boundary value problem: for a given function $f \in S(\Gamma_0)$ find $u \in S(\Omega)$, such that $u|_{\Gamma_0} = f$ and

$$a_\lambda(u,v) = 0, \qquad \forall v \in S_0(\Omega). \tag{2.12.23}$$

It is known that (2.12.23) admits a unique solution for $\lambda = 0$, denoted by u_0. Let $u_1 = u - u_0$, then u_1 satisfies

$$a_0(u_1, v) = -\lambda(u_1 + u_0, v), \qquad \forall v \in S_0(\Omega). \tag{2.12.24}$$

We consider an auxiliary problem: for a given function $p \in L^2(\Omega)$ find $u \in S_0(\Omega)$, such that

$$a_0(u,v) = (p,v), \qquad \forall v \in S_0(\Omega). \tag{2.12.25}$$

(2.12.25) defines a bounded linear operator $U : L^2(\Omega) \to S_0(\Omega)$. By (2.12.24) we have

$$u_1 = -\lambda U(u_1 + u_0).$$

By the imbedding theorem, $U : L^2(\Omega) \to L^2(\Omega)$ is a compact operator. Then applying the Riesz-Schauder Theorem [62], we conclude that $I + \lambda U$ is invertible except at a countable set of isolated eigenvalues. Let $R(\lambda)$ denote the inverse operator, then

$$u_1 = \lambda R(\lambda) U u_0,$$

that is

$$u = u_1 + u_0 = u_0 + \lambda R(\lambda) U u_0 \equiv V(\lambda) f.$$

We have already proved the following:

Theorem 2.12.1. *There is a bounded linear operator $V(\lambda) : S(\Gamma_0) \to S(\Omega)$ which is analytic for all $\lambda \in \mathbb{C}$ except a countable set of points, and (2.12.23) admits a unique solution which can be expressed by $u = V(\lambda) f$, for those λ.*

From Theorem 2.12.1 we discover immediately that the transfer matrix $X(\lambda)$ possesses the same analyticity.

Let μ be the least eigenvalue of the problem (2.12.23), then $V(\lambda)$ is analytic in the circular disc $|\lambda| < \mu$. We consider the behavior of the solution u when a point x tends to the point O. By §1.14 we have

$$y_k = X(\lambda \xi^{2(k-1)}) X(\lambda \xi^{2(k-2)}) \cdots X(\lambda \xi^2) X(\lambda) y_0.$$

$|\lambda \xi^{2(k-1)}|$ can be arbitrary small for sufficiently large k. We are only concerned about the asymptotic behavior of y_k as $k \to \infty$, therefore there is no harm in assuming that $|\lambda| \leq \mu - \varepsilon$, where $\varepsilon > 0$.

Theorem 2.12.2. If $|\lambda| \leq \mu - \varepsilon$, then the product of matrices,

$$X(\lambda\xi^{2(k-1)})X(\lambda\xi^{2(k-2)})\cdots X(\lambda\xi^2)X(\lambda)$$

converges uniformly with respect to λ, as $k \to \infty$.

Proof. We have the Jordan canonical form,

$$TX_0T^{-1} = J_0 = \begin{pmatrix} 1 & \\ & J_1 \end{pmatrix},$$

where $X_0 = X(0)$, and J_1 consists of the Jordan blocks associated with those eigenvalues of the matrix X_0 satisfying $|\lambda| < 1$. Let

$$J(\lambda) = TX(\lambda)T^{-1},$$

then $J(0) = J_0$. $J(\lambda)$ is an analytic function with respect to the independent variable λ. Let

$$Y_k(\lambda) = J(\lambda\xi^{2(k-1)})J(\lambda\xi^{2(k-2)})\cdots J(\lambda\xi^2)J(\lambda). \qquad (2.12.26)$$

It will suffice to prove the uniform convergence of $Y_k(\lambda)$ with respect to λ as $k \to \infty$. (2.3.8) implies that $J_1^k \to 0$ as $k \to \infty$. We take an appropriate natural number m, such that

$$\|J_0^m\| = \left\|\begin{pmatrix} 1 & \\ & J_1^m \end{pmatrix}\right\| = 1.$$

Since $Y_m(\lambda)$ is an analytic function,

$$\|Y_m(\lambda) - J_0^m\| < C|\lambda|. \qquad (2.12.27)$$

Let $\xi^{2m} = \eta$, then $0 < \eta < 1$. By the convergence of infinite product we get

$$\|Y_m(\lambda\eta^k)Y_m(\lambda\eta^{k-1})\cdots Y_m(\lambda)\| \leq \prod_{i=0}^{k}(1 + |\lambda|\eta^i) < C.$$

Let l be the greatest integer part of k/m for $k \geq 1$, then by (2.12.26)

$$Y_k(\lambda) = J(\lambda\xi^{2(k-1)})\cdots J(\lambda\xi^{2ml})Y_m(\lambda\xi^{2m(l-1)})\cdots Y_m(\lambda).$$

In this expression the number of matrices $J(\lambda\xi^{2(k-1)}),\cdots,J(\lambda\xi^{2ml})$ is only finite, therefore $\|Y_k(\lambda)\|$ are uniformly bounded with respect to k and λ. Since $Y_k(\lambda)$ is an analytic function with respect to the independent variable λ, (2.12.27) can be strengthened to

$$\|Y_k(\lambda) - J_0^k\| < C|\lambda|, \quad k = 1, 2, \ldots, \qquad (2.12.28)$$

where the constant C is independent of k and λ.

2.12 NONHOMOGENEOUS AND HELMHOLTZ EQUATIONS

Using the above estimate, we can prove the uniform convergence of $Y_k(\lambda)$ by means of the Cauchy criterion. Let $k, l > m$, then

$$\begin{aligned}\|Y_k(\lambda) - Y_l(\lambda)\| &= \|(Y_{k-m}(\lambda\xi^{2m}) - Y_{l-m}(\lambda\xi^{2m}))Y_m(\lambda)\| \\ &\leq \|Y_{k-m}(\lambda\xi^{2m}) - Y_{l-m}(\lambda\xi^{2m})\| \cdot \|Y_m(\lambda)\| \\ &\leq \Big\{\|Y_{k-m}(\lambda\xi^{2m}) - J_0^{k-m}\| + \|J_0^{k-m} - J_0^{l-m}\| \\ &\quad + \|J_0^{l-m} - Y_{l-m}(\lambda\xi^{2m})\|\Big\} \cdot \|Y_m(\lambda)\| \\ &< C|\lambda|\xi^{2m} + C\|J_0^{k-m} - J_0^{l-m}\|.\end{aligned}$$

J_0^k converges as $k \to \infty$. For an arbitrary $\varepsilon > 0$, we first take m sufficiently large, next take k, l sufficiently large, then the right hand side of the above inequality would be less than ε. □

Theorem 2.12.3. *For any fixed $g \in \mathbb{C}^n$, $X(\lambda\xi^{2(k-1)})X(\lambda\xi^{2(k-2)})\cdots X(\lambda)g \to \alpha(\lambda)g_1$ as $k \to \infty$, where g_1 is the first eigenvector of the matrix X_0, $g_1 = (1, 1, \ldots, 1)^T$, $\alpha(\lambda)$ is an analytic function with respect to the independent variable λ, which depends on g linearly. And the above limit is uniform with respect to λ.*

Proof. Let
$$Z_k(\lambda) = X(\lambda\xi^{2(k-1)})X(\lambda\xi^{2(k-2)})\cdots X(\lambda),$$
$$Z_\infty(\lambda) = \lim_{k\to\infty} Z_k(\lambda).$$

Note that
$$Z_\infty(0) = \lim_{k\to\infty} X_0^k,$$
and denote this limit by X_0^∞, then by the uniformity of convergence and the analyticity of $Z_k(\lambda)$ we get
$$\|Z_\infty(\lambda) - X_0^\infty\| < C|\lambda|.$$

We also have
$$Z_\infty(\lambda) = Z_\infty(\lambda\xi^{2k})Z_k(\lambda),$$
therefore
$$\|Z_\infty(\lambda) - X_0^\infty Z_k(\lambda)\| < C|\lambda|\xi^{2k},$$
thus
$$|Z_\infty(\lambda)g - X_0^\infty Z_k(\lambda)g| < C|\lambda|\xi^{2k}.$$

Applying Lemma 2.2.5 we obtain $X_0^\infty Z_k(\lambda)g = \alpha_k(\lambda)g_1$, where $\alpha_k(\lambda)$ is a constant for fixed k and λ, therefore $\alpha_k(\lambda)$ converges uniformly with respect to λ as $k \to \infty$. Denote by $\alpha(\lambda)$ the limit of $\alpha_k(\lambda)$. $\alpha(\lambda)$ depends on g linearly, because (2.12.23) is a linear problem. □

Now we consider the singularity at the point O of the solution u to (2.12.23). Let $\lambda u = -p$, then applying Lemma 2.12.4 we can construct $z_k^{(m)}$ and $u^{(m)}$, such that $u_2 = \sum_{m=1}^{\infty} u^{(m)}$ is a particular solution to the equation (2.12.1), and satisfies

$$\|u_2\|_{0,\infty,\Omega} + \|u_2\|_{1,\Omega} \leq C|\lambda|.$$

The eigenvalues of the matrix X_0 are arranged in order such that $1 = \lambda_1 > |\lambda_2| \geq |\lambda_3| \geq \cdots$. Let the associated eigenvectors be $g_1, g_2, \ldots$ and so on.

Theorem 2.12.4. *If λ_2 is a simple root and $\lambda_2 > \xi^2$, then the solution to the problem (2.12.23) satisfies*

$$\lim_{k \to \infty} (y_k - \alpha(\lambda)g_1)\lambda_2^{-k} = \beta(\lambda)g_2, \quad (2.12.29)$$

where $\beta(\lambda)$ is an analytic function which depends on f linearly, and the above limit is uniform with respect to λ.

Proof. By (2.12.20) we get

$$z_k^{(m)} = X_0^{k-m} z_m^{(m)}$$

for $k \geq m$. Theorem 2.12.2 implies that u is bounded, and the inequality (2.12.22) gives

$$|z_m^{(m)}| \leq C|\lambda|\xi^{2m}.$$

Theorem 2.3.3 shows that the transfer matrix X_0 admits an eigenvalue $\lambda_1 = 1$ which corresponds to a linear elementary divisor and an eigenvector g_1, therefore there exist analytic functions $\alpha^{(m)}(\lambda)$ and $\beta^{(m)}(\lambda)$, such that

$$\lim_{k \to \infty} (z_k^{(m)} - \alpha^{(m)}(\lambda)g_1)\lambda_2^{m-k} = \beta^{(m)}(\lambda)g_2,$$

where

$$|\alpha^{(m)}(\lambda)| + |\beta^{(m)}(\lambda)| \leq C|\lambda|\xi^{2m}. \quad (2.12.30)$$

Denote by $\tilde{u}$ the interpolating function of $\lambda_2^k g_2$, $k = 0, 1, \ldots$, on Ω, then

$$\|\tilde{u}\|_{0,\infty,\Omega} + \|\tilde{u}\|_{1,\Omega} \leq C. \quad (2.12.31)$$

Let

$$z_k^{*(m)} = z_k^{(m)} - \alpha^{(m)}(\lambda)g_1 - \beta^{(m)}(\lambda)g_2\lambda_2^{k-m}, \quad k = 1, 2, \ldots,$$

and denote by $u^{*(m)}$ the interpolating function of $z_k^{*(m)}$, then by (2.12.22),(2.12.31) and (2.12.30) we have

$$\|u^{*(m)}\|_{0,\infty,\Omega} + \|u^{*(m)}\|_{1,\Omega} \leq C|\lambda|\xi^{2m} + C|\alpha^{(m)}(\lambda)| + C|\beta^{(m)}(\lambda)\lambda_2^{-m}|$$

$$\leq C|\lambda|\left\{\xi^{2m} + \left(\frac{\xi^2}{\lambda_2}\right)^m\right\}.$$

Let $w = u - \sum_{m=1}^{\infty} u^{*(m)}$. It satisfies the homogeneous equation on Ω, and belongs to $S(\Omega)$.

We investigate the singularity of $u - w$ at the point O. From the definition of $z^{*(m)}$ we get

$$|z_k^{*(m)}| \leq \begin{cases} C|\lambda|\xi^{2m}(|\lambda_3| + \varepsilon)^{k-m}, & \text{as } k \geq m, \\ C|\lambda|\xi^{2m}\lambda_2^{k-m}, & \text{as } k < m, \end{cases}$$

where $\varepsilon > 0$ is such a constant that $\xi^2 < |\lambda_3| + \varepsilon < \lambda_2$, then

$$\left|\sum_{m=1}^{\infty} z_k^{*(m)} \lambda_2^{-k}\right| \leq C|\lambda| \sum_{m=1}^{k} \xi^{2m}(|\lambda_3| + \varepsilon)^{k-m}\lambda_2^{-k} + C|\lambda| \sum_{m=k+1}^{\infty} \xi^{2m}\lambda_2^{-m}$$

$$= C|\lambda| \frac{\xi^{2(k+1)} - \xi^2(|\lambda_3| + \varepsilon)^k}{\xi^2 - (|\lambda_3| + \varepsilon)}\lambda_2^{-k} + C|\lambda| \frac{\xi^{2(k+1)}/\lambda_2^{k+1}}{1 - \xi^2/\lambda_2}$$

$$\leq C|\lambda| \max\left\{\left(\frac{\xi^2}{\lambda_2}\right)^k, \left(\frac{|\lambda_3| + \varepsilon}{\lambda_2}\right)^k\right\}.$$

Uniformly with respect to λ

$$\lim_{k \to \infty} \sum_{m=1}^{\infty} z_k^{*(m)} \lambda_2^{-k} = 0,$$

as $k \to \infty$. Hence it has no contribution to the singularity near the point O. The singularity of u is the same as w. $\square$

We conclude this section with proving the inequality (1.14.14).

Theorem 2.12.5. μ, μ_1, μ_2 are all negative numbers, and $-\mu \leq -\mu_1 < -\mu_2$.

Proof. We investigate the Rayleigh ratio

$$R(u) = \frac{\int |\nabla u|^2 \, dx}{\int u^2 \, dx}$$

on $S_0(\Omega_0)$ and $S_0(\Omega)$ separately. We have $R(u) > 0$, and [58]

$$-\mu = \min_{\substack{u \in S_0(\Omega_0) \\ u \neq 0}} R(u), \quad -\mu_1 = \min_{\substack{u \in S_0(\Omega) \\ u \neq 0}} R(u).$$

If $u \in S_0(\Omega)$, we extend it to zero on Ω^*. In this sense $S_0(\Omega) \subset S_0(\Omega_0)$, hence $-\mu \leq -\mu_1$. As for μ_2, by (1.14.9) it is known that $\mu_2 = \frac{\mu_1}{\xi^2}$, thus $-\mu_2 > -\mu_1$, since $0 < \xi < 1$. $\square$

2.13 Discontinuous boundary value problems

Let Ω_0 be a bounded polygonal domain. We consider the Dirichlet problems of the Laplace equation,

$$\triangle u = 0, \quad x \in \Omega_0, \tag{2.13.1}$$

$$u\big|_{\partial \Omega_0} = f, \tag{2.13.2}$$

where f is a piecewise constant function defined on the boundary.

We use infinite element mesh as usual, where an infinite number of elements is applied near every discontinuous point of f. The infinite element space is defined as

$$S(\Omega_0) = \left\{ u \in C(\Omega_0); u\big|_{e_i} \in P_1(e_i), i = 1, 2, \ldots, \sup |u| < +\infty \right\}.$$

Let $\varphi_1, \varphi_2, \ldots, \varphi_i, \ldots$ be shape functions associated with all interior nodes, then the infinite element solution is defined as: $u \in S(\Omega_0), u\big|_{\partial \Omega} = f$, and

$$a(u, \varphi_i) = 0, \quad \forall \varphi_i. \tag{2.13.3}$$

We start with studying uniqueness. Let u_1, u_2 be two solutions to (2.13.3), then $u = u_1 - u_2$ satisfies homogeneous boundary condition.

Lemma 2.13.1. *The difference $u = u_1 - u_2$ belongs to the space $H^1(\Omega_0)$.*

Proof. We consider a corner and assume that the solutions are equal to constants α and β on Γ^* and Γ_*, stated in §1.11. The local subdomain is denoted by Ω and an infinite similar mesh is given on Ω, stated in §1.7. The theory in §2.6 is valid here with minor changes, for instance the transfer matrices X and $\tilde{X}$ are $(n-2) \times (n-2)$ matrices now. We take an integer $N \geq 2$, and consider the domain $\Omega \setminus \overline{\xi^N \Omega}$. Since the eigenvalues of X and $\tilde{X}$ satisfy $|\lambda| < 1$, an analogous result to Theorem 2.6.2 tells us the general solution is

$$y_k = \tilde{X}^k z_1 + X^{N-k} z_2, \quad k = 0, 1, \cdots N, \tag{2.13.4}$$

where $z_1, z_2 \in \mathbb{R}^{n-2}$ are arbitrary. Moreover, let V be the invariant subspace of $\mathbb{R}^{n-2}$ corresponding to the eigenvalues $\lambda \neq 0$ of the matrix X. If $y_k = B_k u$, then $z_2 \in V$. This is because we can replace N by $N+n$ in (2.13.4), then $y_k = \tilde{X}^k z_1 + X^{N+n-k} z_3$ for two vectors z_1 and z_3, and $z_2 = X^n z_3 \in V$.

We replace N by $N+1$ in (2.13.4), then there are two vectots z_4 and z_5, such that

$$y_k = \tilde{X}^k z_4 + X^{N+1-k} z_5, \quad k = 0, 1, \cdots N + 1. \tag{2.13.5}$$

Since the decomposition is unique, comparing (2.13.4) with (2.13.5) we get

$$z_1 = z_4, \quad z_2 = X z_5.$$

The mapping X is invertiable on V, thus $z_5 = X^{-1}z_2$. Therefore $y_{N+1} = \tilde{X}^{N+1}z_1 + X^{-1}z_2$. By induction we have

$$y_k = \tilde{X}^k z_1 + X^{N-k} z_2, \qquad k = N+1, N+2, \cdots.$$

The eigenvalue of $\tilde{X}^k$ satisfy $|\lambda| < 1$, and the eigenvalues of X^{-1} on V satisfy $|\lambda| > 1$. But y_k is bounded, so $z_2 = 0$, and

$$y_k = \tilde{X}^k z_1, \qquad k = 0, 1, \cdots. \tag{2.13.6}$$

Lemma 2.1.2 implies that $u \in H^1(\Omega)$. We consider all discontinuous points and conclude that $u \in H^1(\Omega_0)$. □

Theorem 2.13.1. *There is at most one solution to the problem (2.13.3).*

Proof. By Lemma 2.13.1, $u \in H_0^1(\Omega_0)$. Since $a(\cdot,\cdot)$ is positive definite, $u = 0$. □

We turn now to consider a neighborhood of a corner, and the following problem: given α, β, and $y_0 \in \mathbb{R}^{n-2}$, find $u \in S(\Omega)$, such that $u\big|_{\Gamma^*} = \alpha, u\big|_{\Gamma_*} = \beta, B_0 u = y_0$, and

$$a(u, \varphi_i) = 0, \qquad \forall \varphi_i.$$

Let us now verify the formula (1.11.7).

Lemma 2.13.2. *If the above problem admits a solution for all y_0, then*

$$y_{k+1} = \tilde{X} y_k + g, \qquad k = 0, 1, \ldots, \tag{2.13.7}$$

where the vector g is independent of k.

Proof. By Theorem 2.13.1, the solution is unique. The mapping $y_0 \to y_1$ is linear. Therefore there is a real matrix X_1 and a real vector g, such that

$$y_1 = X_1 y_0 + g. \tag{2.13.8}$$

Let $w(x) = u(\xi^k x)$, then $B_0 w = y_k$, and w is the solution with the boundary data α, β, y_k, hence

$$y_{k+1} = X_1 y_k + g.$$

The matrix X_1 and the vector g is independent of k. Let u_1, u_2 be two solutions corresponding to different y_0, say $y_0^{(1)}$ and $y_0^{(2)}$. If $y_k = B_k(u_1 - u_2)$, then by (2.13.8),

$$y_1 = X_1(y_0^{(1)} - y_0^{(2)}),$$

consequently

$$y_k = X_1^k(y_0^{(1)} - y_0^{(2)}). \tag{2.13.9}$$

Since $u_1 - u_2$ is an infinite element solution with homogeneous boundary conditions on Γ^* and Γ_*, (2.13.9) implies $X_1 = \tilde{X}$. □

Using (2.13.7) we have constructed a solution in §1.11. Let us verify it is indeed the solution to (2.13.3), which proves the existence of a solution.

Theorem 2.13.2. *There is a solution to the problem (2.13.3).*

Proof. We construct the solution u by the approach of §1.11. From (1.11.9) we obtain the vector g, then we have $y_k, k = 0, 1, \ldots$, satisfying (1.11.7). The transfer matrix satisfies (1.11.10), therefore (1.11.8), the equation on Γ_1, is satisfied. By the same reason the equation (2.13.3) is satisfied for all interior nodes. To verify that u is a solution, we need only to show u is a bounded function.

Lemma 2.13.2 implies
$$y_k = \tilde{X}^k y_0 + (\tilde{X}^{k-1} + \tilde{X}^{k-2} + \cdots + I)g.$$

Since the spectral radius of the matrix $\tilde{X}$ is less than 1, the above series converges. Hence $|y_k| \leq C$. The proof is complete. □

2.14 Structure of the solutions near a singular point to interface problems

We consider the problems stated in §1.15. For simplicity, we change some notations. Let $x = (x_1, x_2)$ be a point in the domain Ω_0. Ω_0 is divided into subdomains $\Omega_l, l = 1, \ldots, L$, as shown in Fig. 23. The equation is

$$-\sum_{i,j=1}^{2} \frac{\partial}{\partial x_j}\left(a_{ij}(x)p(x)\frac{\partial u}{\partial x_i}\right) + a_0(x)u = f(x). \tag{2.14.1}$$

We assume that $a_{ij} \in C^1(\bar{\Omega}_0), f \in L^2(\Omega_0), a_0 \in L^\infty(\Omega_0)$, and $p(x)$ is a constant on each subdomain Ω_l.

Denote by $x^{(m)}, m = 1, \ldots, M$, the singular points, and $\Omega^{(m)}$ a neighborhood of $x^{(m)}$ which does not contain any other singular points. Let $\Omega^* = \Omega_0 \setminus \bigcup_{m=1}^{M} \overline{\Omega^{(m)}}$. The bilinear form corresponding to (2.14.1) is

$$a(u,v) = \int_\Omega \left(\sum_{i,j=1}^{2} a_{ij} p \frac{\partial u}{\partial x_i}\frac{\partial v}{\partial x_j} + a_0 uv\right) dx.$$

Let $u \in H_0^1(\Omega_0)$ satisfy

$$a(u,v) = (f,v), \quad \forall v \in H_0^1(\Omega_0). \tag{2.14.2}$$

Theorem 2.14.1. *u belongs to H^2 on each domain $\Omega^* \cap \Omega_l$, and $u = v + w$ on $\Omega^{(m)}$. $v \in H^1(\Omega^{(m)})$ and satisfies the equation*

$$-\sum_{i,j=1}^{2} \frac{\partial}{\partial x_j}\left(a_{ij}(x^{(m)})p(x)\frac{\partial v}{\partial x_i}\right) = 0, \tag{2.14.3}$$

and the boundary condition,
$$v|_{\partial\Omega_0 \cap \overline{\Omega^{(m)}}} = 0.$$
w belongs to H^2 on each domain $\Omega^{(m)} \cap \Omega_l$.

Proof. The results of [54] are applied here. It was proved in [54] that u belongs to H^2 on $\Omega^* \cap \Omega_l$. The following argument is concerning a subdomain $\Omega^{(m)}$.

By [54], there is a one-to-one C^1 mapping $F : x \to y$ for appropriately chosen $\Omega^{(m)}$ with the following features: $\Omega^{(m)}$ is mapped onto a neighborhood of the point O on the y-plane, and $y(x^{(m)}) = 0$. F belongs to C^2 except for the point $x^{(m)}$, and the second order derivatives are uniformly bounded. Each interface and boundary curve starting from $x^{(m)}$ is mapped onto one straight line on the y-plane, $u(x(y)) = v_1(y) + w_1(y), v_1 \in H^1$ is the weak solution to the equation

$$-\nabla(p(x(y))\nabla v_1) = 0, \qquad (2.14.4)$$

and the boundary condition, and $w_1(y)$ belongs to H^2 on each subdomain.

Let $G : x \to y$ be the Frechet derivatives of F at the point $x^{(m)}$. Since G is a linear mapping, it also maps the above interfaces and boundary curves onto straight lines, and the slopes of these two families of straight lines coincide at $y = 0$. Thus these straight lines coincide, and consequently, $\Omega^{(m)} \cap \Omega_l$ is mapped onto an identical domain under the mapping F or G.

By [54], the mapping F transfers the equation (2.14.1) to the equation (2.14.4) on the boundary and interfaces. The mapping G is the linear main part of the above mapping, therefore it is easy to see that it transfers (2.14.3) to (2.14.4) on the entire domain. Thus the function $v_1 \circ G$ satisfies the equation (2.14.3) on $\Omega^{(m)}$. Since $w_1 \circ F$ belongs to H^2 on each subdomain, to prove the theorem, it suffices to verify that $v_1 \circ F - v_1 \circ G$ also belongs to H^2 on each subdomain.

By [54], v_1 is a finite sum of functions with the form $r^\alpha \Theta(\vartheta)$ in polar coordinates, where $\alpha < 1$ and $\Theta(\vartheta)$ are piecewise smooth solutions to the corresponding Sturm-Liouville problem. $v_1 \in H^1$ implies $\alpha > 0$. Let us evaluate the second order derivatives:

$$\frac{\partial^2}{\partial x_i \partial x_j} v_1(F(x)) = \sum_{m,l} \frac{\partial^2 v_1}{\partial y_m \partial y_l} \frac{\partial y_m}{\partial x_i} \frac{\partial y_l}{\partial x_j} + \sum_m \frac{\partial v_1}{\partial y_m} \frac{\partial^2 y_m}{\partial x_i \partial x_j},$$

$$\frac{\partial^2}{\partial x_i \partial x_j} v_1(G(x)) = \sum_{m,l} \frac{\partial^2 v_1}{\partial z_m \partial z_l} \frac{\partial z_m}{\partial x_i} \frac{\partial z_l}{\partial x_j},$$

where $y = F(x), z = G(x)$. F is smooth, hence

$$|y - z| \le Cr^2, \quad \left|\frac{\partial y_m}{\partial x_i} - \frac{\partial z_m}{\partial x_i}\right| \le Cr,$$

and then the expression of v_1 gives

$$\left|\frac{\partial^2 v_1}{\partial y_m \partial y_l}\right| \le Cr^{\alpha-2}, \quad \left|\frac{\partial^2 v_1}{\partial z_m \partial z_l}\right| \le Cr^{\alpha-2},$$

$$\left|\frac{\partial^2 v_1}{\partial y_m \partial y_l} - \frac{\partial^2 v_1}{\partial z_m \partial z_l}\right| \le Cr^{\alpha-3}|y-z|.$$

Therefore
$$\left|\frac{\partial^2}{\partial x_i \partial x_j}v_1(F(x)) - \frac{\partial^2}{\partial x_i \partial x_j}v_1(G(x))\right| \le Cr^{\alpha-1},$$
which belongs to L^2. □

Notes to Chapter 2

The references of this chapter are roughly the same as those of Chapter One. Some new material is added too. The purpose of this chapter is to verify the algorithm given in Chapter One.

The contents of §2.1, §2.2, §2.3, §2.5 are the theoretical basis of this chapter. It has mainly drawn its material from [4], we still narrate it through the Laplace equation as an example here. Lemma 2.1.1, Theorem 2.3.1 and Theorem 2.3.2 are new. The reference failed a little in its rigour, and those three propositions further ensure the theory to be mathematically rigorous.

We consider the plane elasticity problems, which is the very topic of [4], in §2.4. For the sake of theoretical completeness, we set specially a section, §2.10, which states and proves general conclusion for quite general problems, based on [19]. When readers handle different problems, concrete conclusions from the general are needed.

To set §2.6 is also for theoretical completeness. Before this section for the eigenvalue method we always assume that the set of eigenvectors of the transfer matrix is a basis of the space. This restriction has not caused any trouble in real computation, for all problems which we have encountered satisfy this restriction. However this restriction makes theory incomplete. To get rid of this restriction, we made some efforts in [4], but the result was not concise enough. We give in this section a formula effective for all cases. The content of this section is based on [22].

The material of §2.7 is new, which is the comment of [17].

The material of §2.8 is based on [7] and [19]. The material of §2.9 is base on [23] and [31], but Theorem 2.9.2 is new, which serves as a rigorous proof of the three dimensional iterative method of the second type, and is also useful in proving Theorem 2.11.3. The most part of §2.11 is based on [26] and [38]. §2.12 is based on [10] and [20]. §2.13 is new which verifies the algorithm given in §1.11 and §1.12. §2.14 is based on [33] which presents an important theorem applied for solving equations with variable coefficients and nonlinear problems.

3 Convergence

We will present various convergence theorems for the infinite element method in this chapter. They indicate that the precision of the infinite element method for those solutions with singularities is not lower than that of the finite element method for those solutions without singularities. Moreover the precision of the infinite element method is even higher. This fact finds expression in two respects. On the one hand it is included in Theorem 3.3.2, where a weighted L^2-norm estimate, stronger than the normal L^2-norm estimate, is given; on the other hand it will be proved in §3.6 that the infinite element solution can be expanded near the singular points like the exact solution, and each term of the expansion approximates to the corresponding term of the exact solution. The property of converging by terms indicates that the infinite element method is a natural method which gives the solutions with singularities automatically.

For the sake of simplicity and conciseness in exposition, we prove the convergence theorems in §3.1–§3.7 only for a model problem. Let Ω_0 be a bounded plane polygonal domain with an interior angle $\alpha > \pi$, and all other interior angles are not greater than π. We construct Cartesian coordinates (O, x_1, x_2) where the origin is just the vertex of the angle α. Let $x = (x_1, x_2)$ or $y = (y_1, y_2)$ be points in Ω_0. We also construct polar coordinates (O, r, ϑ). Consider the Dirichlet problem of the Laplace equation,

$$\begin{cases} \triangle u = 0, & x \in \Omega_0, \\ u|_{\partial \Omega_0} = f. \end{cases} \tag{3.0.1}$$

We assume that $f = 0$ near the point O, then we have the expansion,

$$u = \sum_{j=1}^{\infty} c_j r^{\frac{j\pi}{\alpha}} \sin \frac{j\pi}{\alpha}(\vartheta - \vartheta_0). \tag{3.0.2}$$

In §3.8 and §3.9 the problems are different, but by the same reason we assume that the elliptic operator in the equations is still the Laplace operator.

As for the convergence for other problems the readers can prove some parallel conclusions themselves, or can consult the references listed at the end of this book.

3.1 Some auxiliary inequalities

In this section we assume that Ω is a bounded domain in $\mathbb{R}^2$ with a Lipschitz continuous boundary.

Theorem 3.1.1. *There is a constant $C > 0$ depending only on Ω, such that for all constants $p \geq 1$ and $u \in H^1(\Omega)$ we have*

$$\|u\|_{0,p,\Omega} \leq C(\|u\|_{0,\Omega} + \sqrt{p}|u|_{1,\Omega}), \tag{3.1.1}$$

$$\|u\|_{0,p,\Gamma} \leq C(\|u\|_{0,\Omega} + \sqrt{p}|u|_{1,\Omega}), \tag{3.1.2}$$

where Γ is a line segment contained in Ω.

Proof. By the Sobolev integral identity [44],

$$u(x) = \sum_{i=0}^{2} w_i(x), \quad \forall x \in \bar{\Omega}, \tag{3.1.3}$$

where

$$w_0(x) = \int_\Omega K(y)u(y)\,dy,$$

$$w_i(x) = \int_\Omega \frac{\partial u(y)}{\partial y_i} \frac{B_i(x,y)}{|x-y|}\,dy, \quad i = 1, 2,$$

and $K(y), B_i(x,y)$ are bounded functions on Ω and $\Omega \times \Omega$ respectively.

By the Schwarz inequality

$$|w_0(x)| \leq C\int_\Omega |u(y)|\,dy \leq C\|u\|_{0,\Omega}.$$

Taking L^2-norm on the both sides we get

$$\|w_0\|_{0,\Omega} \leq C\|u\|_{0,\Omega}.$$

It has no harm in assuming $p > 2$. We set $\rho = |x-y|$, then by the Hölder inequality,

$$|w_i(x)| \leq C \int_\Omega \left|\frac{\partial u}{\partial y_i}\right|^{\frac{2}{p}} \rho^{-\frac{1}{p}} \left|\frac{\partial u}{\partial y_i}\right|^{1-\frac{2}{p}} \rho^{\frac{1}{p}-1}\,dy$$

$$\leq C \left(\int_\Omega \left|\frac{\partial u}{\partial y_i}\right|^2 \rho^{-1}\,dy\right)^{\frac{1}{p}} \left(\int_\Omega \left|\frac{\partial u}{\partial y_i}\right|^2\,dy\right)^{\frac{1}{2}-\frac{1}{p}} \left(\int_\Omega \rho^{\frac{2}{p}-2}\,dy\right)^{\frac{1}{2}}$$

$$\leq C\sqrt{p}|u|_{1,\Omega}^{1-\frac{2}{p}} \left(\int_\Omega \left|\frac{\partial u}{\partial y_i}\right|^2 \rho^{-1}\,dy\right)^{\frac{1}{p}}.$$

Taking L^p-norm on the both sides and changing the order of the integrals on the right, we obtain

$$\|w_i\|_{0,p,\Omega} \leq C\sqrt{p}|u|_{1,\Omega}, \quad i = 1, 2.$$

From (3.1.3) we get (3.1.1). The proof of (3.1.2) is analogous. □

Corollary. *For any measurable set $D \subset \Omega$ or line segment $\Gamma \subset \Omega$ if $d = (\operatorname{meas} D)^{\frac{1}{2}} < \sigma$ or $h = \operatorname{meas} \Gamma < \sigma$, then for all $u \in H^1(\Omega)$ we have*

$$\|u\|_{0,D} \leq Cd |\log d|^{\frac{1}{2}} \|u\|_{1,\Omega}, \tag{3.1.4}$$

$$\|u\|_{0,\Gamma} \leq Ch |\log h|^{\frac{1}{2}} \|u\|_{1,\Omega}, \tag{3.1.5}$$

where the constant $\sigma \in (0,1)$ and the constant C depends only on σ and Ω.

Proof. By the Hölder inequality we have

$$\|u\|_{0,D} \leq (\operatorname{meas} D)^{\frac{1}{2} - \frac{1}{p}} \|u\|_{0,p,\Omega}, \quad \forall p > 2.$$

Then by Theorem 3.1.1 we get

$$\begin{aligned} \|u\|_{0,D} &\leq C(\operatorname{meas} D)^{\frac{1}{2} - \frac{1}{p}} \sqrt{p} \|u\|_{1,\Omega} \\ &= Cd^{1 - \frac{2}{p}} \sqrt{p} \|u\|_{1,\Omega}. \end{aligned}$$

Taking $p = 3|\log \sigma|^{-1}|\log d| > 2$ we get (3.1.4). The proof of (3.1.5) is analogous. □

Let us introduce some terminologies and notations [61]. Let $\hat{D} \subset \mathbb{R}^2$ be a given bounded domain. If for $D \subset \mathbb{R}^2$ there is an invertible affine transformation $F : \hat{x}(\in \mathbb{R}^2) \to F(\hat{x}) = B\hat{x} + b \in \mathbb{R}^2$, such that $D = F(\hat{D})$, then it is called that D and $\hat{D}$ are affine equivalent. The collection of all subsets of Ω affine equivalent to $\hat{D}$ is denoted by $\mathcal{K}(\Omega)$. Let $\hat{P}$ be a finite dimensional function space on $\hat{D}$, $D \in \mathcal{K}(\Omega)$, and we define

$$P_D = \{p : D \to \mathbb{R}; p = \hat{p} \circ F^{-1}, \hat{p} \in \hat{P}\},$$

$$H(D, \Omega) = \{v \in H^1(\Omega); v|_D \in P_D\}.$$

Theorem 3.1.2. *There is a constant $C > 0$, such that for any $D \in \mathcal{K}(\Omega)$, and any $u \in H(D, \Omega)$, we have*

$$\|u\|_{0,\infty,D} \leq C \frac{h_D}{\rho_D} |\log h_D|^{\frac{1}{2}} \|u\|_{1,\Omega}, \tag{3.1.6}$$

provided $h_D = \operatorname{diam} D < 1$, where diam stands for diameter and

$$\rho_D = \sup\{\operatorname{diam} G, G \text{ is a disk contained in } D\}.$$

Proof. If $\hat{u} = u(F(\hat{x})) \in \hat{P}$, $\hat{x} \in \hat{D}$, then

$$\|u\|_{0,\infty,D} = \|\hat{u}\|_{0,\infty,\hat{D}}.$$

Since $\hat{P}$ is finite dimensional, there is a constant $C > 0$, such that

$$\|\hat{u}\|_{0,\infty,\hat{D}} \leq C\|\hat{u}\|_{0,\hat{D}}.$$

But

$$\|\hat{u}\|_{0,\hat{D}} = |\det B|^{-\frac{1}{2}}\|u\|_{0,D},$$

$$|\det B| = \frac{\operatorname{meas} D}{\operatorname{meas} \hat{D}} \geq \frac{\pi}{4} \frac{\rho_D^2}{\operatorname{meas} \hat{D}}.$$

Therefore

$$\|u\|_{0,\infty,D} \leq C\rho_D^{-1}\|u\|_{0,D}.$$

Applying the Corollary of Theorem 3.1.1, we get (3.1.6). □

3.2 Approximate properties of piecewise polynomials

We introduce some weighted Sobolev spaces. Let $\Omega \subset \mathbb{R}^2$ be a bounded domain with Lipschitz continuous boundary. We introduce the following norms and semi-norms in $C^\infty(\bar{\Omega})$ with $p \geq 1, \sigma < 2$,

$$\|u\|_{0(\sigma),\Omega} = \left(\int_\Omega r^{-\sigma}|u|^2\,dx\right)^{\frac{1}{2}},$$

$$|u|_{k,p(\sigma),\Omega} = \left\{\int_\Omega r^{p(k-1)-\sigma}|\partial^k u|^p\,dx\right\}^{\frac{1}{p}},$$

$$\|u\|_{m,p(\sigma),\Omega} = \left\{\|u\|_{0,p,\Omega}^p + \sum_{k=1}^m |u|_{k,p(\sigma),\Omega}^p\right\}^{\frac{1}{p}},$$

where $|\partial^k u| = \sum_{\alpha+\beta=k}\left|\frac{\partial^k u}{\partial x_1^\alpha \partial x_2^\beta}\right|$, $k \geq 1$. Th completion of the space $C^\infty(\bar{\Omega})$ with respect to the above norms are denoted by $H^{0(\sigma)}(\Omega)$ and $W^{m,p(\sigma)}(\Omega)$ respectively. And we denote $H^{m(\sigma)}(\Omega) = W^{m,2(\sigma)}(\Omega), \|\cdot\|_{m(\sigma),\Omega} = \|\cdot\|_{m,2(\sigma),\Omega}$. By (3.0.2) it is easy to see that the solution u to (3.0.1) belongs to $W^{m,p(\sigma)}$ near the point O, where $\sigma < 2 - (1 - \frac{\pi}{\alpha})p$.

Theorem 3.2.1. (a)

$$W^{2,p(\sigma)}(\Omega) \to C(\bar{\Omega}) \quad (p > 1, 2 - p < \sigma < 2),$$

especially

$$H^{2(\sigma)}(\Omega) \to C(\bar{\Omega}) \quad (0 < \sigma < 2);$$

(b)
$$H^1(\Omega) \to H^{0(\sigma)}(\Omega) \quad (0 < \sigma < 2),$$

where " $\to$ " expresses continuous imbedding.

Proof. (a) By definition it will suffice to prove there is a constant $C > 0$, such that
$$\max_{x \in \bar{\Omega}} |u(x)| \leq C\|u\|_{2,p(\sigma),\Omega}, \quad \forall u \in C^\infty(\bar{\Omega}). \tag{3.2.1}$$

Without losing generality we assume that Ω is star-shape with respect to a disk contained in Ω, then we have the Sobolev integral identity [44],

$$u(x) = \int_\Omega u(y)p(y)\,dy + \sum_{i=1}^2 b_i(x) \int_\Omega \frac{\partial u(y)}{\partial y_i} p(y)\,dy$$

$$+ \sum_{|\alpha|=2} \int_\Omega \partial^\alpha u(y) B_\alpha(x,y)\,dy,$$

where α is a multi-index, $p(y)$ is a smooth function, and $b_i(x), B_\alpha(x,y)$ are bounded functions depending only on Ω. By the Hölder inequality we get

$$|u(x)| \leq C\|u\|_{0,p,\Omega} + C\left|\int_\Omega r^{-\frac{\sigma}{p}} |\partial u| r^{\frac{\sigma}{p}}\,dy\right|$$
$$+ C\int_\Omega r^{\frac{p-\sigma}{p}} |\partial^2 u| r^{-\frac{p-\sigma}{p}}\,dy$$
$$\leq C\|u\|_{0,p,\Omega} + C|u|_{1,p(\sigma),\Omega} \left(\int_\Omega r^{\frac{\sigma}{p-1}}\,dy\right)^{\frac{p-1}{p}}$$
$$+ C|u|_{2,p(\sigma),\Omega} \left(\int_\Omega r^{-\frac{p-\sigma}{p-1}}\,dy\right)^{\frac{p-1}{p}}.$$

Since $\sigma > 2 - p$, the integrals
$$\int_\Omega r^{\frac{\sigma}{p-1}}\,dy, \quad \int_\Omega r^{-\frac{p-\sigma}{p-1}}\,dy$$
are finite. Thus
$$|u(x)| \leq C\|u\|_{2,p(\sigma),\Omega},$$
which is just (3.2.1).

(b) We take $p = \frac{\sigma+2}{2\sigma}, q = \frac{2p}{p-1}$, then since $H^1(\Omega) \to L^q(\Omega)$, we get

$$\|u\|_{0(\sigma),\Omega} \leq \|u\|_{0,q,\Omega} \left(\int_\Omega r^{-\sigma p}\,dx\right)^{\frac{1}{p}} \leq C\|u\|_{1,\Omega}.$$

from the Hölder inequality. $\square$

We turn now to the domain Ω_0 mentioned at the beginning of this chapter. Ω_0 is divided into a countably infinite number of triangular elements $e_1, e_2, \ldots, e_i, \ldots$, which is finite away from any neighborhood of the point O. As usual we require that every side of every triangle is either a part of the boundary $\partial\Omega_0$, or a side of another triangle. Denote by h the maximum length of the diameters of all triangular elements. We require in addition the partition satisfies the following conditions:

(A) The partition is regular, that is the interior angles of all elements possess a common lower bound $\vartheta_0 > 0$.

(B) $O \notin \bar{e}_i$, $\forall i \in \mathbb{N}$. And let $d(O, e_i)$ be the distance from e_i to the point O, then there is a constant χ, such that

$$\text{meas}\, e_i \leq \chi h^2 (d(O, e_i))^2, \qquad \forall i \in \mathbb{N}.$$

For instance the similar partition given in Chapter One satisfies the condition (B).

We define infinite element subspaces ($m \geq 1$),

$$S^m(\Omega_0) = \{u \in H^1(\Omega_0); |u|_{e_i} \in P_m(e_i), \forall i \in \mathbb{N}\}.$$

$S^m(\Omega_0)$ is simply denoted by $S(\Omega_0)$ if $m = 1$. We make equipartition on three sides of each triangle where each side is divided into m segments, then we get $3m$ nodes along the boundary of this triangle. Starting from those nodes we construct parallel lines to the sides, and let the crossing points be nodes. Thus $\frac{1}{2}(m+1)(m+2)$ nodes are obtained. Define an interpolation operator $\Pi : C(\bar{\Omega}_0) \to S^m(\Omega_0)$ with respect to these nodes.

Theorem 3.2.2. *If the conditions (A),(B) hold, and $p > 1, 2 - p < \sigma < p, 0 \leq s \leq 1, 1 - s + \frac{\sigma}{p} \geq 0$, then we have the interpolating estimate,*

$$\|u - \Pi u\|_{s,p,\Omega_0} \leq C h^{m+1-s} |u|_{m+1,p(\sigma),\Omega_0}, \quad \forall u \in W^{m+1,p(\sigma)}(\Omega_0). \tag{3.2.2}$$

Proof. We consider the cases of $s = 0, 1$ first. By Theorem 3.2.1, $W^{m+1,p(\sigma)}(\Omega_0) \to C(\bar{\Omega}_0)$, so Πu makes sense. According to the interpolating estimate on triangular elements [61],

$$|u - \Pi u|_{s,p,e_i} \leq C h_i^{m+1-s} |u|_{m+1,p,e_i}, \tag{3.2.3}$$

where h_i is the diameter of e_i.

By the condition (A) we have

$$\text{meas}\, e_i \geq h_i^2 \sin^3 \vartheta_0 \cos \vartheta_0,$$

then by the condition (A),(B) we have

$$h_i^2 \leq \chi \frac{h^2(d(O, e_i))^2}{\sin^3 \vartheta_0 \cos \vartheta_0}.$$

Substituting it into (3.2.3) we get

$$|u - \Pi u|_{s,p,e_i} \leq Ch^{m+1-s}(d(O,e_i))^{m+1-s}|u|_{m+1,p,e_i}.$$

We have $d(O, e_i) \leq r$ on the triangle e_i, therefore

$$(d(O,e_i))^{m+1-s}|u|_{m+1,p,e_i} \leq (d(O,e_i))^{1-s+\frac{\sigma}{p}}|u|_{m+1,p(\sigma),e_i}.$$

By the assumption of the boundedness of the domain and of $1 - s + \frac{\sigma}{p} \geq 0$ we obtain

$$|u - \Pi u|_{s,p,e_i} \leq Ch^{m+1-s}|u|_{m+1,p(\sigma),e_i}.$$

Upon summing them with respect to i we get (3.2.2). If $0 < s < 1$, using the interpolation inequality [55], we also get (3.2.2). □

Theorem 3.2.3. *If the conditions (A),(B) hold, and $u \in W^{1,p}(\Omega_0)$ for $p > 2$, then*

$$\lim_{h \to 0} \|u - \Pi u\|_{1,p,\Omega_0} = 0.$$

Proof. Because $W^{1,p}(\Omega_0) \to C(\bar{\Omega}_0)$ [60], Πu makes sense. Theorem 3.2.2 implies that (3.2.2) holds for the functions in $C^\infty(\bar{\Omega}_0)$. $C^\infty(\bar{\Omega}_0)$ is dense in $W^{1,p}(\Omega_0)$, hence for any $\varepsilon > 0$ we can take $v \in C^\infty(\bar{\Omega}_0)$, such that $\|v - u\|_{1,p} < \varepsilon$. We take h small enough such that $\|v - \Pi v\|_{1,p} < \varepsilon$. Besides, Π is a bounded operator in $W^{1,p}(\Omega_0)$, therefore

$$\|\Pi v - \Pi u\|_{1,p} \leq C\|v - u\|_{1,p}.$$

Combining the above inequalities we obtain the desired result. □

Following the same lines as the proof of Theorem 3.2.2, we get

Theorem 3.2.4. *If the conditions (A),(B) hold, and $p > 1, 2 - p < \sigma < p$, then we have*

$$\|u - \Pi u\|_{0,\infty,\Omega_0} \leq Ch^{m+1-\frac{2}{p}}|u|_{m+1,p(\sigma),\Omega_0}, \quad \forall u \in W^{m+1,p(\sigma)}(\Omega_0).$$

3.3 H^1 and L^2 convergence

Let

$$a(u,v) = \int \nabla u \cdot \nabla v \, dx.$$

If $f \in H^{\frac{1}{2}}(\partial\Omega_0)$, we take $u_0 \in H^1(\Omega_0)$, such that $u_0|_{\partial\Omega_0} = f$. The formulation of the problem (3.0.1) in weak form is: find $u \in H^1(\Omega_0)$, such that $u - u_0 \in H_0^1(\Omega_0)$, and

$$a(u,v) = 0, \quad \forall v \in H_0^1(\Omega_0). \tag{3.3.1}$$

Having made an infinite element partition, we obtain the infinite element space $S^m(\Omega_0)$ and the interpolation operator Π. Let $S_0^m(\Omega_0) = S^m(\Omega_0) \cap H_0^1(\Omega_0)$. We take an approximation to u_0, $u_0^I \in S^m(\Omega_0)$, then the infinite element approximate solution u_h satisfies: $u_h \in S^m(\Omega_0)$, $u_h - u_0^I \in S_0^I \subset S_0^m(\Omega_0)$, and

$$a(u_h, v) = 0, \qquad \forall v \in S_0^m(\Omega_0). \qquad (3.3.2)_m$$

Here the subscript of $(3.3.2)_m$ means that m-th order elements are applied.

We estimate the error $u - u_h$. In accordance with the boundary condition we define a set

$$V = \{v^I \in S^m(\Omega_0); v^I|_{\partial\Omega_0} = (\Pi u - u_0^I)|_{\partial\Omega_0}\}.$$

Theorem 3.3.1. *If the conditions (A),(B) hold, the solution to the problem (3.3.1), $u \in H^{m+1(\sigma)}(\Omega_0)$ $(0 < \sigma < \frac{2\pi}{\alpha})$, and if u_h is the solution to the problem $(3.3.2)_m$, then*

$$\|u - u_h\|_{1,\Omega_0} \leq C(h^m |u|_{m+1(\sigma),\Omega_0} + \min_{v^I \in V} \|v^I\|_{1,\Omega_0}). \qquad (3.3.3)$$

Proof. We take an arbitrary $v^I \in V$, and set $v = v^I - \Pi u + u_h$, then $v \in S_0^m(\Omega_0)$. (3.3.1) and $(3.3.2)_m$ yield

$$a(u - u_h, v) = 0.$$

Thus

$$\begin{aligned}|u - u_h|_{1,\Omega_0}^2 &= a(u - u_h, u - u_h) \\ &= a(u - u_h, u - \Pi u) + a(u - u_h, \Pi u - u_h) \\ &= a(u - u_h, u - \Pi u) + a(u - u_h, v^I).\end{aligned}$$

Applying the Schwarz inequality we get

$$|u - u_h|_{1,\Omega_0} \leq |u - \Pi u|_{1,\Omega_0} + |v^I|_{1,\Omega_0}. \qquad (3.3.4)$$

Then applying the Friedrichs inequality and the trace theorem we obtain

$$\begin{aligned}\|u - u_h\|_{0,\Omega_0} &\leq C(|u - u_h|_{1,\Omega_0} + \|u - u_0^I\|_{0,\partial\Omega_0}) \\ &\leq C(|u - u_h|_{1,\Omega_0} + \|u - \Pi u\|_{0,\partial\Omega_0} + \|\Pi u - u_0^I\|_{0,\partial\Omega_0}) \\ &\leq C(|u - u_h|_{1,\Omega_0} + \|u - \Pi u\|_{1,\Omega_0} + \|v^I\|_{1,\Omega_0}).\end{aligned}$$

By (3.3.4) we have

$$\|u - u_h\|_{1,\Omega_0} \leq C(\|u - \Pi u\|_{1,\Omega_0} + \|v^I\|_{1,\Omega_0}).$$

Substituting the estimate in Theorem 3.2.2 into it and taking the minimum with respect to v^I, we get (3.3.3). □

Corollary. If $u_0^I = \Pi u$, then

$$\|u - u_h\|_{1,\Omega_0} \le Ch^m |u|_{m+1(\sigma),\Omega_0}.$$

The above corollary shows that the error estimate of the infinite element method is optimal.

We turn now to the weighted L^2-norm estimate. Let us first consider an auxiliary problem,

$$\begin{cases} -\triangle \varphi = g, & x \in \Omega_0, \\ \varphi|_{\partial \Omega_0} = 0. \end{cases} \quad (3.3.5)$$

We take $\sigma \in (0, \frac{2\pi}{\alpha})$ and set $\tilde{\sigma} = \sigma - 2$. By Theorem 3.2.1, if $g \in H^{0(\tilde{\sigma})}(\Omega_0)$, then $\int_{\Omega_0} gv\, dx$ defines a continuous linear functional on $H_0^1(\Omega_0)$. Hence the problem (3.3.5) admits a unique weak solution $\varphi \in H_0^1(\Omega_0)$. Moreover the following conclusion is valid [50]:

Lemma 3.3.1. *The weak solution φ to the problem (3.3.5) belongs to $H^{2(\sigma)}(\Omega_0)$, and there exists a constant $C > 0$ independent of g, such that*

$$\|\varphi\|_{2(\sigma),\Omega_0} \le C\|g\|_{0(\tilde{\sigma}),\Omega_0}.$$

Let Ω be a polygon with L sides, and $l_j (j = 1, \ldots, L)$ be the sides of Ω. Using Aubin-Nitsche's trick [61], we can prove the following:

Theorem 3.3.2. *Under the hypotheses of Theorem 3.3.1 if we require in addition that u_0, u_0^I vanish on the adjacent sides to the point O, then we have*

$$\|u - u_h\|_{0(-\tilde{\sigma}),\Omega_0}$$
$$\le C\left(h^{m+1}|u|_{m+1(\sigma),\Omega_0} + h \min_{v^I \in V} \|v^I\|_{1,\Omega_0} + \sum_{j=1}^L \|u_0 - u_0^I\|_{-\frac{1}{2}, l_j}\right). \quad (3.3.6)$$

Proof. Let $g = r^{\tilde{\sigma}}(u - u_h)$. Because $u - u_h \in H^1(\Omega_0)$, by Theorem 3.2.1 we know $g \in H^{0(\tilde{\sigma})}(\Omega_0)$. Let φ be the solution to the problem (3.3.5) with the above function g, then by Lemma 3.3.1, $\varphi \in H^{2(\sigma)}(\Omega_0)$, and

$$\|\varphi\|_{2(\sigma),\Omega_0} \le C\|u - u_h\|_{0(-\tilde{\sigma}),\Omega_0}. \quad (3.3.7)$$

Theorem 3.2.1 implies that $\varphi \in C(\bar{\Omega}_0)$, therefore the interpolating function $\Pi \varphi$ makes sense. By Theorem 3.2.2 we have

$$\|\varphi - \Pi\varphi\|_{1,\Omega_0} \le Ch|\varphi|_{2(\sigma),\Omega_0}. \quad (3.3.8)$$

Applying the Green's formula and (3.3.1),(3.3.2)$_m$ we obtain

$$\|u-u_h\|^2_{0(-\tilde{\sigma}),\Omega_0} = \int_{\Omega_0} r^{\tilde{\sigma}}(u-u_h)(u-u_h)\,dx$$

$$= \int_{\Omega_0}(-\Delta\varphi)(u-u_h)\,dx = a(u-u_h,\varphi) - \int_{\partial\Omega_0}(u-u_h)\frac{\partial\varphi}{\partial\nu}\,ds$$

$$= a(u-u_h,\varphi-\Pi\varphi) - \int_{\partial\Omega_0}(u_0-u_0^I)\frac{\partial\varphi}{\partial\nu}\,ds. \qquad (3.3.9)$$

Noting that u_0 and u_0^I vanish on the adjacent sides to the point O, we have

$$\left|\int_{\partial\Omega_0}(u_0-u_0^I)\frac{\partial\varphi}{\partial\nu}\,ds\right| \leq \sum_{j=1}^{L}\|u_0-u_0^I\|_{-\frac{1}{2},l_j}\left\|\frac{\partial\varphi}{\partial\nu}\right\|_{\frac{1}{2},l_j}$$

$$\leq C\sum_{j=1}^{L}\|u_0-u_0^I\|_{-\frac{1}{2},l_j}\|\varphi\|_{2(\sigma),\Omega_0}.$$

By (3.3.8) we get

$$|a(u-u_h,\varphi-\Pi\varphi)| \leq |u-u_h|_{1,\Omega_0}|\varphi-\Pi\varphi|_{1,\Omega_0}$$
$$\leq Ch|u-u_h|_{1,\Omega_0}|\varphi|_{2(\sigma),\Omega_0}.$$

Upon substituting them into (3.3.9) we get

$$\|u-u_h\|^2_{0(-\tilde{\sigma}),\Omega_0} \leq C\left(h|u-u_h|_{1,\Omega_0} + \sum_{j=1}^{L}\|u_0-u_0^i\|_{-\frac{1}{2},l_j}\right)\|\varphi\|_{2(\sigma),\Omega_0}.$$

Then (3.3.7) gives

$$\|u-u_h\|_{0(-\tilde{\sigma}),\Omega_0} \leq C\left(h|u-u_h|_{1,\Omega_0} + \sum_{j=1}^{L}\|u_0-u_0^I\|_{-\frac{1}{2},l_j}\right). \qquad (3.3.10)$$

Finally using Theorem 3.3.1 we obtain (3.3.6). □

Corollary. If $u_0^I = \Pi u$, then

$$\|u-u_h\|_{0(-\tilde{\sigma}),\Omega_0} \leq Ch^{m+1}\|u\|_{m+1(\sigma),\Omega_0}. \qquad (3.3.11)$$

Proof. Using the one dimensional interpolating estimate we have [56]

$$\|u-\Pi u\|_{-\frac{1}{2},l_j} \leq Ch^{m+1}\|u\|_{m+\frac{1}{2},l_j}.$$

provided l_j is not an adjacent side to the point O. Then applyihg the trace theorem we get

$$\|u\|_{m+\frac{1}{2},l_j} \leq C\|u\|_{m+1(\sigma),\Omega_0}.$$

Substituting them into (3.3.6) leads to (3.3.11). □

3.4 Maximum principle and uniform convergence

Starting from this section we consuder linear elements only, that is we consider infinite element solutions in the space $S(\Omega)$ only. Denote by b_i $(i = 1, 2, \dots)$ the nodes on $\bar{\Omega}_0$. The shape functions $\varphi_1, \varphi_2, \dots$ satisfy: $\varphi_i \in S(\Omega_0)$, and

$$\varphi_i(b_j) = \delta_{ij}, \qquad i,j = 1, 2, \cdots. \tag{3.4.1}$$

The set of finite linear combinations of $\varphi_i's$,

$$\varphi = c_1\varphi_1 + c_2\varphi_2 + \cdots + c_j\varphi_j,$$

is denoted by $\tilde{S}(\Omega_0)$. Then each $\varphi \in \tilde{S}(\Omega_0)$ satisfies $O \notin \operatorname{supp} \varphi$. Concerning the infinite element partition we require it to satisfy, besides the conditions (A),(B) in §3.2, two more conditions:

(C) As subspaces in $H^1(\Omega_0)$, $\tilde{S}(\Omega_0)$ is dense in $S(\Omega_0)$.

(D) The interior angles of all elements e_i are not greater than $\frac{\pi}{2}$.

We have proved in §2.3 that the similar partition constructed in Chapter One satisfies the condition (C).

To prove the maximum principle, we prove some auxiliary lemmas first [48,57].

Lemma 3.4.1. *Let* $a_{ij} = a(\varphi_i, \varphi_j)$ *and assume that the condition* (D) *holds. then*

$$a_{ij} \leq 0, \qquad i \neq j, b_i, b_j \in \bar{\Omega}_0. \tag{3.4.2}$$

Proof. If b_i, b_j are not adjacent to each other, then $a_{ij} = 0$. If they are adjacent, then they belong to one same element or two. Let $\lambda_1, \lambda_2, \lambda_3$ be barycentric coordinates on the element e_k, then by the condition (D) we have

$$\int_{e_k} \nabla \lambda_1 \cdot \nabla \lambda_2 \, dx \leq 0,$$

λ_2 and λ_3, or λ_3 and λ_1 possess the same property. Therefore (3.4.2) holds. □

Lemma 3.4.2. *We assume the conditions* (C),(D) *hold. If* $w \in S(\Omega_0)$, α *is a constant, and* $w_\alpha \in S(\Omega_0)$ *is determined by*

$$w_\alpha(b_i) = \min\{\alpha, w(b_i)\}, \quad i = 1, 2, \ldots, \tag{3.4.3}$$

then $v_\alpha = w - w_\alpha \in S(\Omega_0)$ *satisfies*

$$a(v_\alpha, v_\alpha) \leq a(w, v_\alpha). \tag{3.4.4}$$

Proof. Because

$$a(w, v_\alpha) = a(v_\alpha, v_\alpha) + a(w_\alpha, v_\alpha),$$

it will suffice to prove

$$a(w_\alpha, v_\alpha) \geq 0.$$

Now because

$$a(\alpha, v_\alpha) = 0,$$

it will suffice to prove

$$a(\alpha - w_\alpha, v_\alpha) \leq 0. \tag{3.4.5}$$

By the condition (C) for any $\varepsilon > 0$ we can take finite sum to approximate $\alpha - w_\alpha$ and v_α as follows:

$$\left\| (\alpha - w_\alpha) - \sum_{i=1}^{N} c_i \varphi_i \right\|_{1,\Omega_0} < \varepsilon,$$

$$\left\| v_\alpha - \sum_{j=1}^{M} d_j \varphi_j \right\|_{1,\Omega_0} < \varepsilon.$$

Since $\alpha - w_\alpha \geq 0, v_\alpha \geq 0, \varphi_i \geq 0$, and c_i, d_i are values at nodes, it is always possible to make $c_i \geq 0$ and $d_i \geq 0$. By Lemma 3.4.1 we have

$$a\left(\sum_{i=1}^{N} c_i \varphi_i, \sum_{j=1}^{M} d_j \varphi_j \right) = \sum_{i,j} a_{ij} c_i d_j \leq 0.$$

But ε is arbitrary, hence we obtain (3.4.5). □

Lemma 3.4.3. *For any* $p \geq 1$ *and any triangular element* e_i, *let* $w \in P_1(e_i)$, *then we have*

$$\sum_{j=1}^{3} |w(b_j)|^p \operatorname{meas} e_i \leq C \|w\|_{0,p,e_i}^p, \tag{3.4.6}$$

where b_1, b_2, b_3 *are vertices on* e_i.

Proof. We define an affine transformation $x = F(y)$ which maps e_i to $\{y = (y_1, y_2) \in \mathbb{R}^2; 0 < y_1 < 1, 0 < y_2 < 1 - y_1\}$. Let $\hat{w}(y) = w(F(y))$, then

$$\|w\|_{0,p,e_i}^p = \int_{e_i} |w(x)|^p \, dx = 2 \operatorname{meas} e_i \int_0^1 \int_0^{1-y_1} |\hat{w}(y)|^p \, dy_2 dy_1.$$

It has no harm in assuming that $|\hat{w}(0)|$ is the maximum absolute value, then $\hat{w}(y)$ has the same sign as $\hat{w}(0)$ on the triangle $\hat{e} = \{y \in \mathbb{R}^2; 0 < y_1 < \frac{1}{2}, 0 < y_2 < \frac{1}{2} - y_1\}$. We define a linear function λ_0 on $\hat{e}$, such that $\lambda_0(0,0) = 1, \lambda_0(\frac{1}{2},0) = \lambda_0(0,\frac{1}{2}) = 0$, then $|\hat{w}(y)| \geq |\hat{w}(0)\lambda_0(y)|$ on $\hat{e}$. Thus

$$\|w\|_{0,p,e_i}^p \geq 2\operatorname{meas} e_i \int_{\hat{e}} |\hat{w}(0)\lambda_0(y)|^p \, dy$$
$$\geq \frac{\operatorname{meas} e_i}{C} |\hat{w}(0)|^p,$$

that is

$$|\hat{w}(0)|^p \operatorname{meas} e_i \leq C\|w\|_{0,p,e_i}^p.$$

Noting that $|\hat{w}(0)|$ is the maximum absolute value, we get (3.4.6). □

Lemma 3.4.4. *If $\varphi(t)$ is a nonnegative and nonincreasing function defined on $[\alpha_0, +\infty)$, and there are constants $C > 0, \gamma > 0, s > 1$, such that*

$$\varphi(\beta) \leq \frac{C}{(\beta - \alpha)^\gamma} (\varphi(\alpha))^s, \qquad (3.4.7)$$

for $\beta > \alpha \geq \alpha_0$, then

$$\varphi(\alpha_0 + d) = 0,$$

where

$$d^\gamma = C(\varphi(\alpha_0))^{s-1} 2^{\frac{\gamma s}{s-1}}.$$

Proof. We take an arbitrary $\alpha \in [\alpha_0, +\infty)$, and take β satisfying $\beta > \alpha$ and

$$(\beta - \alpha)^\gamma \geq C 2^{\frac{\gamma}{s-1}} (\varphi(\alpha))^{s-1},$$

then by (3.4.7) we have

$$\varphi(\beta) \leq 2^{-\frac{\gamma}{s-1}} \varphi(\alpha). \qquad (3.4.8)$$

We construct an increasing sequence $\alpha_0, \alpha_1, \ldots, \alpha_n, \ldots$, satisfying a recurrent relation,

$$(\alpha_n - \alpha_{n-1})^\gamma = C 2^{\frac{\gamma}{s-1}} (\varphi(\alpha_{n-1}))^{s-1},$$

then we have

$$(\alpha_n - \alpha_{n-1})^\gamma \leq C 2^{\frac{\gamma}{s-1}} (2^{-\frac{\gamma}{s-1}} \varphi(\alpha_{n-2}))^{s-1}$$
$$= 2^{-\gamma} (\alpha_{n-1} - \alpha_{n-2})^\gamma,$$

that is

$$\alpha_n - \alpha_{n-1} \leq \frac{1}{2}(\alpha_{n-1} - \alpha_{n-2}). \qquad (3.4.9)$$

Let $n \to \infty$, then by (3.4.8) we get

$$\lim_{n \to \infty} \varphi(\alpha_n) = 0.$$

By (3.4.9) we have
$$\lim_{n\to\infty} \alpha_n \leq \alpha_0 + 2(\alpha_1 - \alpha_0) = \alpha_0 + d.$$

Therefore
$$\varphi(\alpha_0 + d) \leq \varphi(\lim_{n\to\infty} \alpha_n) = 0.$$

But by the assumption $\varphi \geq 0$, hence $\varphi(\alpha_0 + d) = 0$. □

We turn now to prove the maximum principle and the uniform convergence of solutions.

Theorem 3.4.1. *If the conditions* (C),(D) *hold and* u_0^I *is continuous on* $\partial\Omega_0$, *then the solution* u_h *to the problem* $(3.3.2)_1$ *always achieves its maximum and minimum value on the boundary* $\partial\Omega_0$.

Proof. Let $\alpha = \max_{x \in \partial\Omega_0} u_0^I(x)$, and let $w_\alpha \in S(\Omega_0)$ satisfy
$$w_\alpha(b_i) = \min\{\alpha, u_h(b_i)\}, \quad i = 1, 2, \ldots,$$

and set $v_\alpha = u_h - w_\alpha$. Then by Lemma 3.4.2, we have
$$a(v_\alpha, v_\alpha) \leq a(u_h, v_\alpha).$$

On the boundary $\partial\Omega_0$ $w_\alpha = u_h$, therefore $v_\alpha \in S_0(\Omega_0)$. By $(3.3.2)_1$ we have
$$a(v_\alpha, v_\alpha) \leq 0.$$

Thus $v_\alpha = 0$, that is $w_\alpha = u_h$, that is
$$u_h(x) \leq \alpha, \quad \forall x \in \Omega_0.$$

Analogously we can get the lower bound of u_h. □

Corollary. *Under the hypotheses of Theorem 3.4.1 the transfer matrix* $X = (x_{ij})$ *satisfies*
$$x_{ij} \geq 0, \quad \forall i, j = 1, 2, \ldots, n.$$

Proof. If it were not the case, then there would be at least a $x_{ij} < 0$. We could take boundary data $u_0^I \geq 0$ appropriately, such that u_h would be negative at one node, which contradicts to Theorem 3.4.1. □

Theorem 3.4.2. *We assume that the conditions* (A),(B),(C),(D) *hold, and* u *is the solution to the problem* (3.3.1), u_h *is the solution to the problem* $(3.3.2)_1$, *where* $u_0^I = \Pi u_0$.

(a) *If* $u \in W^{1,p}(\Omega_0), 2 < p < \frac{2}{1-\frac{\pi}{\alpha}}$, *then*
$$\lim_{h\to 0} \|u - u_h\|_{0,\infty,\Omega_0} = 0, \qquad (3.4.10)$$

3.4 MAXIMUM PRINCIPLE

(b) If $u \in W^{2,p(\sigma)}(\Omega_0), 2 < p < \frac{2}{1-\frac{\pi}{\alpha}}, 0 \leq \sigma < 2 - (1 - \frac{\pi}{\alpha})p$, then

$$\|u - u_h\|_{0,\infty,\Omega_0} \leq Ch\|u\|_{2,p(\sigma),\Omega_0}. \tag{3.4.11}$$

Proof. By the embedding theorem [60], $W^{1,p}(\Omega_0) \to C(\bar{\Omega}_0)$, and by Theorem 3.2.1 $W^{2,p(\sigma)}(\Omega_0) \to C(\bar{\Omega}_0)$, hence the interpolating function Πu makes sense. By $(3.3.1), (3.3.2)_1$ we have

$$a(u_h - \Pi u, v) = a(u - \Pi u, v), \quad \forall v \in S_0(\Omega_0).$$

Let $w = u_h - \Pi u \in S_0(\Omega_0), f_i = \frac{\partial(u-\Pi u)}{\partial x_i}, i = 1, 2$, then we have

$$a(w, v) = \sum_{i=1}^{2} \int_{\Omega_0} f_i \frac{\partial v}{\partial x_i} dx.$$

We take an arbitrary $\alpha \geq 0$, and define $w_\alpha \in S_0(\Omega_0)$, satisfying

$$w_\alpha(b_i) = \min\{\alpha, w(b_i)\}, \quad i = 1, 2, \cdots.$$

Then by Lemma 3.4.2, $v_\alpha = w - w_\alpha$ satisfies

$$a(v_\alpha, v_\alpha) \leq a(w, v_\alpha) = \sum_{i=1}^{2} \int_{\Omega_0} f_i \frac{\partial v_\alpha}{\partial x_i} dx.$$

We take a constant q, such that $\frac{1}{p} + \frac{1}{q} = 1$, then by the Hölder inequality we get

$$a(v_\alpha, v_\alpha) \leq \sum_{i=1}^{2} \|f_i\|_{0,p,\Omega_0} \left\|\frac{\partial v_\alpha}{\partial x_i}\right\|_{0,q,\Omega_0}. \tag{3.4.12}$$

Let $E(\alpha) = \{x \in \Omega_0; v_\alpha(x) > 0\}$. Since $v_\alpha \geq 0$, $\overline{E(\alpha)}$ is the union of some triangular elements, and $v_\alpha = 0$ on $\Omega_0 \setminus \overline{E(\alpha)}$. Applying the Hölder inequality we obtain

$$\left\|\frac{\partial v_\alpha}{\partial x_i}\right\|_{0,q,\Omega_0} = \left\|\frac{\partial v_\alpha}{\partial x_i}\right\|_{0,q,E(\alpha)} \leq \left\|\frac{\partial v_\alpha}{\partial x_i}\right\|_{0,E(\alpha)} (\text{meas } E(\alpha))^{\frac{1}{2}-\frac{1}{p}}$$
$$\leq |v_\alpha|_{1,\Omega_0}(\text{meas } E(\alpha))^{\frac{1}{2}-\frac{1}{p}}.$$

Substituting it into (3.4.12) we get

$$|v_\alpha|_{1,\Omega_0}^2 \leq \sum_{i=1}^{2} \|f_i\|_{0,p,\Omega_0} |v_\alpha|_{1,\Omega_0} (\text{meas } E(\alpha))^{\frac{1}{2}-\frac{1}{p}}.$$

We take $\gamma > (\frac{1}{2} - \frac{1}{p})^{-1}$, then by the embedding theorem $H^1(\Omega_0) \to L^\gamma(\Omega_0)$, that is

$$\|v_\alpha\|_{0,\gamma,\Omega_0} \leq C\|v_\alpha\|_{1,\Omega_0}.$$

We notice that $v_\alpha \in S_0(\Omega_0)$, then by the Friedrichs inequality we get

$$\|v_\alpha\|_{0,\gamma,\Omega_0} \leq C|v_\alpha|_{1,\Omega_0}.$$

Hence

$$\|v_\alpha\|_{0,\gamma,\Omega_0} \leq \sum_{i=1}^{2} \|f_i\|_{0,p,\Omega_0} (\text{meas } E(\alpha))^{\frac{1}{2}-\frac{1}{p}}.$$

Now we assume that $\beta > \alpha$ is another real number. $E(\beta)$ is defined on the analogy of $E(\alpha)$. By Lemma 3.4.3 we have

$$\|v_\alpha\|_{0,\gamma,\Omega_0}^\gamma = \sum_{e_i \subset \overline{E(\alpha)}} \int_{e_i} (v_\alpha(x))^\gamma \, dx$$

$$\geq \frac{1}{C} \sum_{b_i \in E(\alpha)} (v_\alpha(b_i))^\gamma \, \text{meas}(\text{supp } \varphi_i)$$

$$\geq \frac{1}{C} \sum_{b_i \in E(\beta)} (v_\alpha(b_i))^\gamma \, \text{meas}(\text{supp } \varphi_i)$$

$$\geq \frac{1}{C}(\beta - \alpha)^\gamma \sum_{b_i \in E(\beta)} \text{meas}(\text{supp } \varphi_i)$$

$$\geq \frac{1}{C}(\beta - \alpha)^\gamma \, \text{meas } E(\beta).$$

Let

$$C_1 = C \left(\sum_{i=1}^{2} \|f_i\|_{0,p,\Omega_0} \right)^\gamma,$$

$$s = \gamma \left(\frac{1}{2} - \frac{1}{p} \right) > 1,$$

$$\varphi(t) = \text{meas } E(t),$$

then we have

$$\varphi(\beta) \leq \frac{C_1}{(\beta - \alpha)^\gamma} (\varphi(\alpha))^s.$$

By Lemma 3.4.4 we have $\varphi(\tilde{\alpha}) = 0$, where

$$\tilde{\alpha} = 2^{\frac{s}{s-1}} [C_1(\varphi(0))^{s-1}]^{\frac{1}{\gamma}}.$$

According to the definition of $\varphi(t)$,

$$v_{\tilde{\alpha}}(x) = 0, \qquad \forall x \in \Omega_0,$$

that is
$$w_{\tilde{\alpha}}(b_i) = \min\{\tilde{\alpha}, w(b_i)\} = w(b_i),$$
therefore
$$w(b_i) \leq \tilde{\alpha} \leq C \sum_{i=1}^{2} \|f_i\|_{0,p,\Omega_0}.$$

Because w is a piecewise linear function,
$$w(x) \leq C \sum_{i=1}^{2} \|f_i\|_{0,p,\Omega_0}, \quad \forall x \in \Omega_0.$$

The same argument is applied to $-w$, then finally we get
$$\|w\|_{0,\infty,\Omega_0} \leq C \sum_{i=1}^{2} \|f_i\|_{0,p,\Omega_0},$$
that is
$$\|u_h - \Pi u\|_{0,\infty,\Omega_0} \leq C|u - \Pi u|_{1,p,\Omega_0}.$$

By the embedding theorem we obtain
$$\begin{aligned}\|u - u_h\|_{0,\infty,\Omega_0} &\leq \|u - \Pi u\|_{0,\infty,\Omega_0} + \|\Pi u - u_h\|_{0,\infty,\Omega_0}\\ &\leq C\|u - \Pi u\|_{1,p,\Omega_0}.\end{aligned}$$

Then Theorem 3.2.2 and Theorem 3.2.3 yield (3.4.10) and (3.4.11). □

3.5 A superconvergence estimate

Starting from this section we assume that the partition is similar near the point O. According to the manner of partition stated in Chapter One, the domain Ω_0 is divided into polygonal domains Ω^* and Ω. Convensional finite element partition is made on Ω^*. $O \in \bar{\Omega}$, the domain Ω is star-shape with respect to the point O, and all interior angles are less than π except the one at the point O. We take a constant $\xi \in (0,1)$, and draw the similar figures of $\partial\Omega$ with centre O and the constants of proportionality $\xi, \xi^2, \cdots, \xi^k, \cdots$. The domain between two polygons forms one "layer". A finite number of rays are drawn from the origin which further divide each layer into a finite number of quadrilaterals. Finally each quadrilateral is divided into two triangles such that the manner of partition for all layers are the same. Under this partition the condition (B) is valid automatically. Denote by h_k the maximum diameter of the elements of the k-th layer, then $h_1 \leq h$, $h_k = \xi^{k-1} h_1$.

In this section we assume that the partition on Ω is more particular, namely the partition satisfies the following condition:

(E) Ω is divided into a finite number of triangular domains $D_1, D_2, \ldots, D_M$ (Fig. 28) by a finite number of rays. In the process of refinement $D_1, D_2, \ldots, D_M$ keep fixed. The triangulation on a given domain D_j along the same direction, and the difference in length among the sub-segments of the line segment AB is at most $O(h^2)$ (Fig.29).

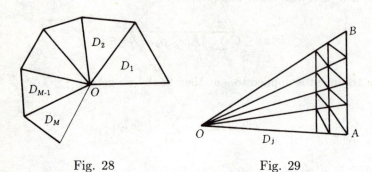

Fig. 28   Fig. 29

For simplicity we assume $\Omega = \Omega_0$. For the general case the superconvergence of the finite element method on Ω^* has been well studied. The readers can consult for example [64].

We introduce some notations first. The rays passing through the point O divide the domain D_j into m "sub-triangles" (Fig.29), denoted by $K_1, \ldots, K_m$. Denote by T_k the k-th layer in D_j, then we have

$$\bar{D}_j = \bigcup_{i=1}^{m} \bar{K}_i = \bigcup_{k=1}^{\infty} \bar{T}_k.$$

Let $e = \triangle b_1 b_2 b_3$, $e' = \triangle b_2 b_4 b_3$ be two triangular elements on $T_k \bigcup T_{k+1}$ (Fig.30). Denote by $(x_1^{(i)}, x_2^{(i)})$ the coordinates of the point b_i. We define the following parameters for e,

$$\xi_1 = x_1^{(2)} - x_1^{(3)}, \quad \xi_2 = x_1^{(3)} - x_1^{(1)}, \quad \xi_3 = x_1^{(1)} - x_1^{(2)},$$
$$\eta_1 = x_2^{(2)} - x_2^{(3)}, \quad \eta_2 = x_2^{(3)} - x_2^{(1)}, \quad \eta_3 = x_2^{(1)} - x_2^{(2)},$$
$$\delta^2 x_1 = x_1^{(1)} - x_1^{(2)} - x_1^{(3)} + x_1^{(4)}, \quad \delta^2 x_2 = x_2^{(1)} - x_2^{(2)} - x_2^{(3)} + x_2^{(4)},$$
$$S = \operatorname{meas} e.$$

Let the local numbering for the nodes on the element e' be $1', 2', 3'$, the geometric parameters of which can be expressed in terms of those of e as follows,

$$\xi_1' = -\xi_1, \quad \xi_2' = -\xi_2 - \delta^2 x_1, \quad \xi_3' = -\xi_3 + \delta^2 x_1,$$
$$\eta_1' = -\eta_1, \quad \eta_2' = -\eta_2 - \delta^2 x_2, \quad \eta_3' = -\eta_3 + \delta^2 x_2.$$

Besides we denote $S' = \operatorname{meas} e'$.

We start with proving some auxiliary lemmas [64].

3.5 A SUPERCONVERGENCE ESTIMATE

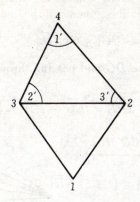

Fig. 30

Lemma 3.5.1. *If the conditions* (A),(E) *hold, then* $e \cup e'$ *forms an "approximate parallelogram", that is*

$$|\overrightarrow{b_1b_2} - \overrightarrow{b_3b_4}| + |\overrightarrow{b_1b_3} - \overrightarrow{b_2b_4}| \leq C\xi^k h^2, \tag{3.5.1}$$

$$|\delta^2 x_1| + |\delta^2 x_2| \leq C\xi^k h^2, \tag{3.5.2}$$

$$|S' - S| \leq C(S + S')h. \tag{3.5.3}$$

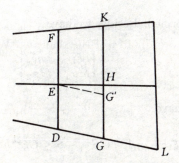

Fig. 31

Proof. See Fig. 31, and we get from the condition (E) that

$$|\overrightarrow{DE} - \overrightarrow{EF}| \leq \xi^k \cdot Ch^2 = C\xi^k h^2. \tag{3.5.4}$$

According to the manner to take the parameter of similarity, ξ, we know

$$|\overrightarrow{GL} - \overrightarrow{DG}| = (1 - \xi)|\overrightarrow{GL}|,$$

Fig. 29 shows that
$$(1-\xi)|\overrightarrow{OA}| \le h,$$
hence
$$|\overrightarrow{GL} - \overrightarrow{DG}| \le Ch|\overrightarrow{GL}| \le C\xi^k h^2. \tag{3.5.5}$$

We draw a line EG' parallel to DG and passing through the point E, then we have
$$|\overrightarrow{DG} - \overrightarrow{EH}| = |\overrightarrow{EG'} - \overrightarrow{EH}| = |\overrightarrow{HG'}|.$$

Since $\triangle OGH$ is similar to $\triangle EG'H$, we have
$$|\overrightarrow{HG'}| = (1-\xi)|\overrightarrow{GH}| \le (1-\xi)\xi^k h,$$
hence
$$|\overrightarrow{DG} - \overrightarrow{EH}| \le C\xi^k h^2. \tag{3.5.6}$$

Besides we have
$$|\overrightarrow{GH} - \overrightarrow{DE}| = |\overrightarrow{HG'}| \le C\xi^k h^2. \tag{3.5.7}$$

Having got the fundamental estimates (3.5.4)–(3.5.7), we can estimate the difference in length of the opposite sides of all quadrilaterals. For instance
$$\overrightarrow{DH} - \overrightarrow{EK} = \overrightarrow{DE} - \overrightarrow{HK} = (\overrightarrow{DE} - \overrightarrow{EF}) + (\overrightarrow{EF} - \overrightarrow{HK}).$$

Thus (3.5.1),(3.5.2) are verified. Besides we have
$$S' - S = \frac{1}{2}\begin{vmatrix} \xi_1 & \eta_1 \\ \delta^2 x_1 & \delta^2 x_2 \end{vmatrix},$$

By the condition (A) we have
$$|\xi_1| + |\eta_1| \le CS^{\frac{1}{2}}, \quad S \le C\xi^{2k}h^2,$$

Noting the inequality (3.5.2) we get (3.5.3). □

Denote by Q the quadrilateral domain shown in Fig. 30. Let $v \in C(Q)$, and v be linear functions on e and e' respectively. Denote by $v^{(1)}, v^{(2)}, v^{(3)}, v^{(4)}$ the nodal values of v, and by $\partial_i v, \partial'_i v, i = 1, 2$, the first order derivatives of v on e and e' respectively, and let
$$\delta^2 v = v^{(1)} - v^{(2)} - v^{(3)} + v^{(4)}.$$

3.5 A SUPERCONVERGENCE ESTIMATE

Lemma 3.5.2. *If the conditions* (A),(E) *hold, and* $v \in S(\Omega)$, *then*

$$\partial_1' v - \partial_1 v = -\eta_1 \frac{\delta^2 v}{S + S'} + O(\xi^{-k})|v|_{1,Q}, \qquad (3.5.8)$$

$$\partial_2' v - \partial_2 v = \xi_1 \frac{\delta^2 v}{S + S'} + O(\xi^{-k})|v|_{1,Q}. \qquad (3.5.9)$$

Proof. We have

$$2S\partial_1 v = \eta_2(v^{(2)} - v^{(1)}) + \eta_3(v^{(3)} - v^{(1)}),$$

$$-2S\partial_2 v = \xi_2(v^{(2)} - v^{(1)}) + \xi_3(v^{(3)} - v^{(1)}),$$

$$v^{(i)} - v^{(j)} = (x_1^{(i)} - x_1^{(j)})\partial_1 v + (x_2^{(i)} - x_2^{(j)})\partial_2 v, \quad 1 \le i, j \le 3,$$

on e. Therefore

$$|v^{(i)} - v^{(j)}| \le \xi^k h |\nabla v| \le C|v|_{1,e}. \qquad (3.5.10)$$

Analogously we have

$$2S'\partial_1' v = -\eta_2(v^{(3)} - v^{(4)}) - \eta_3(v^{(2)} - v^{(4)}) + \delta^2 x_2(v^{(2)} - v^{(3)}),$$

$$-2S'\partial_2' v = -\xi_2(v^{(3)} - v^{(4)}) - \xi_3(v^{(2)} - v^{(4)}) + \delta^2 x_1(v^{(2)} - v^{(3)}),$$

on e'. Upon subtracting, they lead to

$$\begin{aligned}2S'\partial_1' v - 2S\partial_1 v &= \eta_2 \delta^2 v + \eta_3 \delta^2 v + \delta^2 x_2(v^{(2)} - v^{(3)}) \\ &= -\eta_1 \delta^2 v + \delta^2 x_2(v^{(2)} - v^{(3)}),\end{aligned}$$

that is

$$(S + S')(\partial_1' v - \partial_1 v) = -\eta_1 \delta^2 v + \delta^2 x_2(v^{(2)} - v^{(3)}) - (S' - S)(\partial_1' v + \partial_1 v),$$

namely

$$\partial_1' v - \partial_1 v = -\eta_1 \frac{\delta^2 v}{S + S'} + \delta^2 x_2 \frac{v^{(2)} - v^{(3)}}{S + S'} - \frac{S' - S}{S + S'}(\partial_1' v + \partial_1 v).$$

Noting that

$$|\partial_1' v + \partial_1 v| \le C(\xi^{-k} h^{-1})|v|_{1,Q},$$

we obtain (3.5.8) from Lemma 3.5.1 and (3.5.10). The proof of (3.5.9) follows the same lines. □

Lemma 3.5.3. If the conditions (A),(E) hold, then we have

$$\left| \int_{b_1 b_2} f(x)\, dx_1 - \int_{b_3 b_4} f(x)\, dx_1 \right| \leq C \int_Q \left(|\nabla f| + \frac{1}{r}|f| \right) dx,$$

for $f \in H^1(Q)$, which is also true if in the curvilinear integral dx_1 is changed to dx_2.

Proof. We make a bilinear transformation which maps Q to a square $\{\zeta = (\zeta_1, \zeta_2); -1 < \zeta_1, \zeta_2 < 1\}$. Then by the conditions (A),(E) and Lemma 3.5.1 we have

$$\det \frac{\partial x}{\partial \zeta} \geq \frac{1}{C} \xi^{2k} h^2,$$

$$\left| \frac{\partial x_i}{\partial \zeta_j} \right| \leq C \xi^k h, \quad 1 \leq i, j \leq 2,$$

$$\left| \frac{\partial^2 x_i}{\partial \zeta_1 \partial \zeta_2} \right| \leq C \xi^k h^2, \quad i = 1, 2.$$

Therefore

$$\left| \int_{b_1 b_2} f(x)\, dx_1 - \int_{b_3 b_4} f(x)\, dx_1 \right| = \left| \int_{-1}^{1} (f(x(\zeta)) \frac{\partial x_1}{\partial \zeta_1} \Big|_{\zeta_2 = -1}^{\zeta_2 = 1} d\zeta_1 \right|$$

$$\leq \int_{-1}^{1} \int_{-1}^{1} \left| \frac{\partial}{\partial \zeta_2} \left(f(x(\zeta)) \frac{\partial x_1}{\partial \zeta_1} \right) \right| d\zeta$$

$$\leq C \int_{-1}^{1} \int_{-1}^{1} (\xi^{2k} h^2 |\nabla f| + \xi^k h^2 |f|)\, d\zeta$$

$$\leq C \int_Q \left(|\nabla f| + \frac{|f|}{\xi^k} \right) dx$$

$$\leq C \int_Q \left(|\nabla f| + \frac{|f|}{r} \right) dx.$$

□

Let h_i be the diamater of the element e_i, then we have

Lemma 3.5.4. If the condition (A) holds, $u \in H^3(e_i)$, and the interpolation operator $\Pi : C(\bar{e}_i) \to P_1(e_i)$ is given, then

$$u(x) - \Pi u(x) = \sum_{j=1}^{3} R_j(x) + \bar{R}(x),$$

$$R_j(x) = -\frac{1}{2}(\xi_j^2 \partial_1^2 u + 2\xi_j \eta_j \partial_1 \partial_2 u + \eta_j^2 \partial_2^2 u) \frac{\lambda_1 \lambda_2 \lambda_3}{\lambda_j},$$

3.5 A SUPERCONVERGENCE ESTIMATE

$$\bar{R}(x) = \frac{1}{2} \int_0^1 t \sum_{j=1}^3 \lambda_j D_j^3 u(tx + (1-t)b_j) \, dt,$$

where $\partial_k = \frac{\partial}{\partial x_k}$, $k = 1, 2$; $D_j = (x_1 - x_1^{(j)})\partial_1 + (x_2 - x_2^{(j)})\partial_2$, $j = 1, 2, 3$, and we have the estimates

$$\|R_j\|_{0,e_i} + h_i |R_j|_{1,e_i} \leq C h_i^2 \|u\|_{3,e_i},$$

$$\|\bar{R}\|_{0,e_i} + h_i |\bar{R}|_{1,e_i} \leq C h_i^3 \|u\|_{3,e_i}.$$

The proof of this lemma is routine. The readers can consult for instance [64].

On the analogy of $S(\Omega_0)$ we can define infinite element spaces $S(D_j)$ (see Fig. 29), and set

$$S_0(D_j) = \{u \in S(D_j); u|_{AB} = 0\},$$

and

$$a(u,v)_{D_j} = \int_{D_j} \nabla u \cdot \nabla v \, dx.$$

We have the following fundamental lemma:

Lemma 3.5.5. *If the conditions* (A),(E) *hold,* $u \in H^{4(\sigma)}(D_j)$, *and the interpolation operator* $\Pi : C(\bar{D}_j) \to S(D_j)$, *is given, then*

$$|a(u - \Pi u, v)_{D_j}| \leq C h^2 |\log h|^\gamma \|u\|_{4(\sigma), D_j} |v|_{1, D_j}, \quad \forall v \in S_0(D_j), \tag{3.5.11}$$

for $0 < h \leq \frac{1}{2}$, *where* $1 < \sigma < \frac{2\pi}{\alpha}, \gamma = \frac{1}{2}$ *for* $\alpha < 2\pi$, *and* $0 < \sigma < 1, \gamma = 1$ *for* $\alpha = 2\pi$.

Proof. For the sake of writing convenience we omit the subscript j of D_j. By Lemma 3.5.4 we have

$$a(u - \Pi u, v)_D = \sum_{j=1}^3 A_j + a(\bar{R}, v)_D,$$

where $A_j = a(R_j, v)_D$. The following estimate proceeds in four steps.

Step 1. Estimate for $a(\bar{R}, v)_D$.

By Lemma 3.5.4 we have

$$|a(\bar{R}, v)_D| \leq \sum_{e_i \subset D} \|\nabla \bar{R}\|_{0,e_i} |v|_{1,e_i}$$

$$\leq \sum_{e_i \subset D} C h_i^2 \|u\|_{3,e_i} |v|_{1,e_i}.$$

Making use of the similarity we get

$$h_i \leq h \xi^{k-1}, \quad \xi^k \leq Cr,$$

for $e_i \subset T_k$, hence

$$|a(\bar{R}, v)_D| \leq \sum_{e_i \subset D} Ch^2 \|u\|_{3(\sigma), e_i} |v|_{1, e_i}$$
$$\leq Ch^2 \|u\|_{3(\sigma), D} |v|_{1, D}.$$

Step 2. Estimate for A_1.

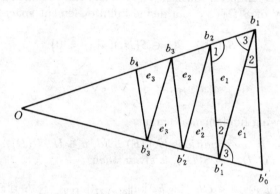

Fig. 32

Take any $K_i \subset D$ and set

$$A_{1\,i} = a(R_1, v)_{K_i},$$

Numbering the elements and nodes in K_i as Fig. 32, by the Green's formula we get

$$A_{1\,i} = \sum_{k=1}^{\infty} \left(\oint_{\partial e_k} + \oint_{\partial e'_k} \right) R_1 (\partial_1 v \, dx_2 - \partial_2 v \, dx_1).$$

R_1 does not vanish on $b_k b'_k$ only, $k = 1, 2, \ldots$, and on these common sides $\lambda'_2 = \lambda_3, \xi'_1 = -\xi_1, \eta'_1 = -\eta_1, R'_1 = R_1$, thus by Lemma 3.5.2 we get

$$A_{1\,i} = B_1 + B_2,$$

where

$$B_1 = \sum_{k=1}^{\infty} O(\xi^{-k}) |v|_{1, e_k \cup e'_k} \int_{b_k b'_k} |R_1| \, ds, \qquad (3.5.12)$$

$$B_2 = -\sum_{k=1}^{\infty} \delta^2 v \int_{b_k b'_k} \frac{R_1}{S_k + S'_k} (\eta_1 \, dx_2 + \xi_1 \, dx_1), \qquad (3.5.13)$$

where S_k and S_k' are the areas of the triangles e_k and e_k' respectively.

Let us estimate B_1 first. By Lemma 3.5.4 and the Corollary of Theorem 3.1.1 we have

$$\int_{b_k b_k'} |R_1|\, ds \leq C h_k^{\frac{5}{2}} \left(\int_{b_k b_k'} |\partial^2 u|^2 \, ds \right)^{\frac{1}{2}} \leq C h_k^{\frac{5}{2}} \xi^{\frac{\sigma-4}{2}k} \left(\int_{b_k b_k'} |r^{\frac{4-\sigma}{2}} \partial^2 u|^2 \, ds \right)^{\frac{1}{2}}$$

$$\leq C h_k^{\frac{5}{2}} \xi^{\frac{\sigma-4}{2}k} h_k^{\frac{1}{2}} |\log h_k|^{\frac{1}{2}} \|u\|_{3(\sigma),D}$$

$$\leq C h^3 \xi^{\frac{2+\sigma}{2}k} |\log h_k|^{\frac{1}{2}} \|u\|_{3(\sigma),D}. \tag{3.5.14}$$

For any constant $\delta > 0$ we have

$$\sum_{k=1}^{\infty} \xi^{\delta k} |\log h_k| = \sum_{k=1}^{\infty} \xi^{\delta k} |\log h| + \sum_{k=1}^{\infty} \xi^{\delta k} k |\log \xi|,$$

and it is easy to see that

$$\sum_{k=1}^{\infty} \xi^{\delta k} = \frac{\xi^\delta}{1 - \xi^\delta} = O(h^{-1}),$$

$$\sum_{k=1}^{\infty} \xi^{\delta k} k = O(h^{-1}),$$

hence

$$\sum_{k=1}^{\infty} \xi^{\delta k} |\log h_k| \leq C h^{-1} |\log h|. \tag{3.5.15}$$

Upon substituting (3.5.14),(3.5.15) into (3.5.12) we get

$$|B_1| \leq C h^3 \|u\|_{3(\sigma),D} \sum_{k=1}^{\infty} |v|_{1, e_k \cup e_k'} \xi^{\frac{\sigma}{2}k} |\log h_k|^{\frac{1}{2}}$$

$$\leq C h^3 \|u\|_{3(\sigma),D} \left(\sum_{k=1}^{\infty} |v|_{1, e_k \cup e_k'}^2 \right)^{\frac{1}{2}} \left(\sum_{k=1}^{\infty} \xi^{\sigma k} |\log h_k| \right)^{\frac{1}{2}}$$

$$\leq C h^{\frac{5}{2}} |\log h|^{\frac{1}{2}} \|u\|_{3(\sigma),D} |v|_{1, K_i}.$$

To estimate B_2, we denote

$$f_k = \frac{R_1}{S_k + S_k'} (\eta_1 \, dx_2 + \xi_1 \, dx_1) \Big|_{b_k b_k'},$$

then
$$B_2 = -\sum_{k=1}^{\infty}(v(b_{k+1}) - v(b'_k) - v(b_k) + v(b'_{k-1}))\int_{b_k b'_k} f_k$$
$$= (v(b_1) - v(b'_0))\int_{b_1 b'_1} f_1$$
$$+ \sum_{k=1}^{\infty}(v(b_{k+1}) - v(b'_k))\left(\int_{b_{k+1} b'_{k+1}} f_{k+1} - \int_{b_k b'_k} f_k\right).$$

We know $v \in S_0(D)$, so the first term of the right hand side of the above expression vanishes. Let us estimate the second term.

$$\left|\int_{b_{k+1} b'_{k+1}} f_{k+1} - \int_{b_k b'_k} f_k\right|$$
$$= \left|\int_{b_{k+1} b'_{k+1}} \frac{\frac{1}{2}(\xi_1^2 \partial_1^2 u + 2\xi_1 \eta_1 \partial_1 \partial_2 u + \eta_1^2 \partial_2^2 u)\lambda_2 \lambda_3}{S_{k+1} + S'_{k+1}}(\eta_1 \, dx_2 + \xi_1 \, dx_1)\right.$$
$$\left. - \int_{b_k b'_k} \frac{\frac{1}{2}(\xi_1^2 \partial_1^2 u + 2\xi_1 \eta_1 \partial_1 \partial_2 u + \eta_1^2 \partial_2^2 u)\lambda_2 \lambda_3}{S_k + S'_k}(\eta_1 \, dx_2 + \xi_1 \, dx_1)\right|.$$

We take one term in it as an example. By Lemma 3.5.1 we have

$$I = \left|\int_{b_{k+1} b'_{k+1}} \frac{\xi_1^3 \partial_1^2 u}{S_{k+1} + S'_{k+1}} dx_1 - \int_{b_k b'_k} \frac{\xi_1^3 \partial_1^2 u}{S_k + S'_k} dx_1\right|$$
$$\leq C\xi^k h^2 \int_{b_k b'_k} |\partial_1^2 u| \, ds + \left|\frac{\xi_1^3}{S_{k+1} + S'_{k+1}}\left(\int_{b_{k+1} b'_{k+1}} \partial_1^2 u \, dx_1 - \int_{b_k b'_k} \partial_1^2 u \, dx_1\right)\right|.$$

Then by Lemma 3.5.3 we have

$$I \leq C\xi^k h^2 \int_{b_k b'_k} |\partial_1^2 u| \, ds + C\xi^k h \int_{e_k \cup e'_{k+1}} \left(|\partial^3 u| + \frac{|\partial^2 u|}{r}\right) dx.$$

In like manner we can estimate the other terms, thus

$$\left|\int_{b_{k+1} b'_{k+1}} f_{k+1} - \int_{b_k b'_k} f_k\right|$$
$$\leq C\xi^k h^2 \int_{b_k b'_k} |\partial^2 u| \, ds + C\xi^k h \int_{e_k \cup e'_{k+1}} \left(|\partial^3 u| + \frac{|\partial^2 u|}{r}\right) dx.$$

On the analogy of (3.5.14) we get

$$\xi^k h^2 \int_{b_k b'_k} |\partial^2 u| \, ds \leq Ch^3 \xi^{\frac{\sigma}{2}k} |\log h_k|^{\frac{1}{2}} \|u\|_{3(\sigma), D},$$

but

$$\xi^k h \int_{e_k \cup e'_{k+1}} \left(|\partial^3 u| + \frac{|\partial^2 u|}{r} \right) dx$$

$$\leq C\xi^{2k} h^2 \xi^{-\frac{4-\sigma}{2}k} \left(\int_{e_k \cup e'_{k+1}} (r^{4-\sigma}|\partial^3 u|^2 + r^{2-\sigma}|\partial^2 u|^2) \, dx \right)^{\frac{1}{2}}$$

$$\leq Ch^2 \xi^{\frac{\sigma}{2}k} \|u\|_{3(\sigma), e_k \cup e'_{k+1}}.$$

Thus

$$|B_2| \leq Ch^3 \|u\|_{3(\sigma),D} \sum_{k=1}^{\infty} \xi^{\frac{\sigma}{2}k} |\log h_k|^{\frac{1}{2}} |v|_{1, e_k \cup e'_k}$$

$$+ Ch^2 \sum_{k=1}^{\infty} \xi^{\frac{\sigma}{2}k} \|u\|_{3(\sigma), e_k \cup e'_{k+1}} \|v\|_{1, e_k \cup e'_k}$$

$$\leq Ch^{\frac{5}{2}} |\log h|^{\frac{1}{2}} \|u\|_{3(\sigma),D} |v|_{1,K_i} + Ch^2 \|u\|_{3(\sigma),K_i} |v|_{1,K_i}. \quad (3.5.16)$$

Combining the above inequalities we obtain

$$|A_1| \leq \sum_{i=1}^{m} |A_{1i}|$$

$$\leq Ch^{\frac{5}{2}} |\log h|^{\frac{1}{2}} \|u\|_{3(\sigma),D} \sum_{i=1}^{m} |v|_{1,K_i} + Ch^2 \sum_{i=1}^{m} \|u\|_{3(\sigma),K_i} |v|_{1,K_i}$$

$$\leq Ch^2 |\log h|^{\frac{1}{2}} \|u\|_{3(\sigma),D} |v|_{1,D}.$$

Step 3. Estimate for A_2.

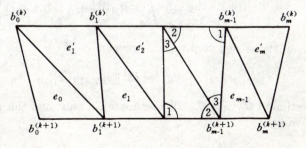

Fig. 33

Fig.33 shows the figure of the k-th layer T_k and some assigned notations, then

$$A_2 = \sum_{k=1}^{\infty} A_{2,k},$$

$$A_{2,k} = a(R_2, v)_{T_k}.$$

On the analogy of the second step, by the Green's formula we have

$$A_{2,k} = \oint_{\partial e_0} R_2(\partial_1 v \, dx_2 - \partial_2 v \, dx_1) + \oint_{\partial e'_m} R_2(\partial_1 v \, dx_2 - \partial_2 v \, dx_1)$$
$$+ \sum_{i=1}^{m-1} \left(\oint_{\partial e_i} + \oint_{\partial e'_i} \right) R_2(\partial_1 v \, dx_2 - \partial_2 v \, dx_1).$$

R_2 does not vanish on $b_i^{(k+1)} b_i^{(k)}$ only, $i = 0, 1, \ldots, m$. Thus on the analogy of the second step we get

$$A_2 = U_1 + U_2 + G_0 + G_m,$$

where

$$U_1 = \sum_{k=1}^{\infty} \sum_{i=1}^{m-2} (v(b_{i+1}^{(k+1)}) - v(b_i^{(k)})) \left(\int_{b_{i+1}^{(k+1)} b_{i+1}^{(k)}} f_{i+1} - \int_{b_i^{(k+1)} b_i^{(k)}} f_i \right),$$

$$U_2 = \sum_{k=1}^{\infty} \sum_{i=1}^{m-2} O(\xi^{-k}) \int_{b_i^{(k+1)} b_i^{(k)}} R_2 |v|_{1, e_i \cup e'_i} \, ds,$$

$$G_0 = \sum_{k=1}^{\infty} \left\{ (v(b_1^{(k+1)}) - v(b_0^{(k)})) \int_{b_1^{(k+1)} b_0^{(k)}} f_1 + \oint_{\partial e_0} R_2(\partial_1 v \, dx_2 - \partial_2 v \, dx_1) \right\},$$

$$G_m = \sum_{k=1}^{\infty} \left\{ (v(b_m^{(k+1)}) - v(b_{m-1}^{(k)})) \int_{b_m^{(k+1)} b_{m-1}^{(k)}} f_{m-1} \right.$$
$$\left. + \oint_{\partial e'_m} R_2(\partial_1 v \, dx_2 - \partial_2 v \, dx_1) \right\}.$$

On the analogy of the second step we can prove

$$|U_1| + |U_2| \leq Ch^2 |\log h|^{\frac{1}{2}} \|u\|_{3(\sigma), D} |v|_{1, D}.$$

It remains to estimate G_0 and G_m. We estimate G_0 only, and the estimate for G_m is analogous. We have

$$2S_0 \partial_1 v = \eta_1 v(b_0^{(k+1)}) + \eta_2 v(b_1^{(k+1)}) + \eta_3 v(b_0^{(k)})$$
$$= \eta_1 (v(b_0^{(k+1)}) - v(b_0^{(k)})) + \eta_2 (v(b_1^{(k+1)}) - v(b_0^{(k)})),$$

3.5 A SUPERCONVERGENCE ESTIMATE

$$-2S_0\partial_2 v = \xi_1 v(b_0^{(k+1)}) + \xi_2 v(b_1^{(k+1)}) + \xi_3 v(b_0^{(k)})$$
$$= \xi_1(v(b_0^{(k+1)}) - v(b_0^{(k)})) + \xi_2(v(b_1^{(k+1)}) - v(b_0^{(k)}))$$

on e_0. Thus

$$\oint_{\partial e_0} R_2(\partial_1 v\, dx_2 - \partial_2 v\, dx_1)$$
$$= -\frac{v(b_0^{(k+1)}) - v(b_0^{(k)})}{2S_0} \int_{b_0^{(k+1)}b_0^{(k)}} R_2(\eta_1\, dx_2 - \xi_1\, dx_1)$$
$$- \frac{v(b_1^{(k+1)}) - v(b_0^{(k)})}{2S_0} \int_{b_0^{(k+1)}b_0^{(k)}} R_2(\eta_2\, dx_2 - \xi_2\, dx_1),$$

besides

$$\int_{b_1^{(k+1)}b_0^{(k)}} f_1 = \frac{1}{S_1 + S_1'} \int_{b_1^{(k+1)}b_1^{(k)}} R_2(\eta_2\, dx_2 - \xi_2\, dx_1),$$

where S_0, S_1, S_1' are the areas of e_0, e_1, e_1' respectively. On the analogy of (3.5.16) we get

$$\left| \sum_{k=1}^{\infty}(v(b_1^{(k+1)}) - v(b_0^{(k)}))\left\{ \frac{1}{2S_0}\int_{b_0^{(k+1)}b_0^{(k)}} R_2(\eta_2\, dx_2 - \xi_2\, dx_1) \right.\right.$$
$$\left.\left. - \frac{1}{S_1 + S_1'}\int_{b_1^{(k+1)}b_1^{(k)}} R_2(\eta_2\, dx_2 - \xi_2\, dx_1) \right\}\right|$$
$$\leq Ch^2|\log h|^{\frac{1}{2}}\|u\|_{3(\sigma),D}|v|_{1,D}.$$

Therefore it remains to estimate

$$I = \sum_{k=1}^{\infty} \frac{v(b_0^{(k+1)}) - v(b_0^{(k)})}{2S_0} \int_{b_0^{(k+1)}b_0^{(k)}} R_2(\eta_1\, dx_2 - \xi_1\, dx_1).$$

The line segments $b_0^{(k+1)}b_0^{(k)}, k = 1, 2, \ldots,$ are situated on one line. It has no harm in assuming the line is just the x_1 axis. Thus

$$I = \sum_{k=1}^{\infty} \frac{v(b_0^{(k+1)}) - v(b_0^{(k)})}{S_0} \int_{b_0^{(k+1)}b_0^{(k)}} (\xi_2^2 \partial_1^2 u + 2\xi_2\eta_2 \partial_1\partial_2 u + \eta_2^2 \partial_2^2 u)\lambda_1\lambda_3\xi_1\, dx_1.$$

We estimate the term with $\partial_1^2 u$, and the estimate for other terms is the same. Noting $\frac{\xi_2^2 \xi_1}{S_0}\frac{1}{\xi^k h} < C$, we only need to estimate

$$I_1 = \sum_{k=1}^{\infty} \int_{\xi^k x_0}^{\xi^{k-1} x_0} \xi^k h(v_{k+1} - v_k)\partial_1^2 u(x_1)\lambda_1\lambda_3\, dx_1,$$

where $x_0 = |\overrightarrow{OB}|$ (see Fig.29), and $v_k = v(b_0^{(k)})$. Noting $v_1 = 0$ we get

$$I_1 = \sum_{k=1}^{\infty} \int_{\xi^k x_0}^{\xi^{k-1} x_0} \xi^k h v_{k+1} \partial_1^2 u(x_1) \lambda_1 \lambda_3 \, dx_1$$

$$- \sum_{k=1}^{\infty} \int_{\xi^{k+1} x_0}^{\xi^k x_0} \xi^k h v_{k+1} \partial_1^2 u(x_1) \lambda_1 \lambda_3 \, dx_1$$

$$= \sum_{k=1}^{\infty} \int_{\xi^k x_0}^{\xi^{k-1} x_0} \xi^k h v_{k+1} \partial_1^2 u(x_1) \lambda_1 \lambda_3 \, dx_1$$

$$- \sum_{k=1}^{\infty} \int_{\xi^k x_0}^{\xi^{k-1} x_0} \xi^k h v_{k+1} \partial_1^2 u(\xi x_1) \lambda_1 \lambda_3 \, d(\xi x_1)$$

$$= \sum_{k=1}^{\infty} \int_{\xi^k x_0}^{\xi^{k-1} x_0} \xi^k h v_{k+1} \int_{\xi}^{1} \frac{\partial}{\partial t}(t \partial_1^2 u(t x_1)) \lambda_1 \lambda_3 \, dt dx_1$$

$$= \int_{\xi}^{1} \sum_{k=1}^{\infty} \int_{\xi^k x_0}^{\xi^{k-1} x_0} \xi^k h v_{k+1} (\partial_1^2 u(t x_1) + t x_1 \partial_1^3 u(t x_1)) \lambda_1 \lambda_3 \, dx_1 dt.$$

Let $t x_1 = \tau$, then we have

$$I_1 = \int_{\xi}^{1} \sum_{k=1}^{\infty} \xi^k h v_{k+1} \int_{t\xi^k x_0}^{t\xi^{k-1} x_0} (\partial_1^2 u(\tau) + \tau \partial_1^3 u(\tau)) \lambda_1 \lambda_3 \, \frac{d\tau}{t} dt.$$

By Theorem 3.1.2 we have

$$|v_{k+1}| \leq C |\log h_k|^{\frac{1}{2}} |v|_{1,D}.$$

Besides we know $1 - \xi \leq Ch$, so by the Schwarz inequality we get

$$|I_1| \leq C \int_{\xi}^{1} \sum_{k=1}^{\infty} \xi^k h |\log h_k|^{\frac{1}{2}} |v|_{1,D} \int_{t\xi^k x_0}^{t\xi^{k-1} x_0} (|\partial_1^2 u(\tau)|$$

$$+ \xi^k |\partial_1^3 u(\tau)|) \, d\tau dt$$

$$\leq C h |v|_{1,D} \int_{\xi}^{1} \sum_{k=1}^{\infty} \xi^k |\log h_k|^{\frac{1}{2}} \cdot \xi^{\frac{k}{2}} h^{\frac{1}{2}} \left(\int_{t\xi^k x_0}^{t\xi^{k-1} x_0} (|\partial_1^2 u(\tau)|^2 \right.$$

$$\left. + \xi^{2k} |\partial_1^3 u(\tau)|^2) \, d\tau \right)^{\frac{1}{2}} dt$$

$$\leq C h^{\frac{3}{2}} |v|_{1,D} \int_{\xi}^{1} \sum_{k=1}^{\infty} \xi^{\frac{\sigma-1}{2} k} |\log h_k|^{\frac{1}{2}} \left(\int_{t\xi^k x_0}^{t\xi^{k-1} x_0} r^{4-\sigma} (|\partial_1^2 u(\tau)|^2 \right.$$

$$\left. + r^2 |\partial_1^3 u(\tau)|^2) \, d\tau \right)^{\frac{1}{2}} dt.$$

If $\alpha < 2\pi$, we take $1 < \sigma < \frac{2\pi}{\alpha}$, then by the Schwarz inequality and (3.5.15) we obtain

$$|I_1| \leq Ch^{\frac{3}{2}}|v|_{1,D}\int_\xi^1\left(\sum_{k=1}^\infty \xi^{(\sigma-1)k}|\log h_k|\right)^{\frac{1}{2}}\left(\sum_{k=1}^\infty \int_{t\xi^k x_0}^{t\xi^{k-1}x_0} r^{4-\sigma}\right.$$

$$\left.\cdot(|\partial_1^2 u(\tau)|^2 + r^2|\partial_1^3 u(\tau)|^2)\,d\tau\right)^{\frac{1}{2}} dt$$

$$\leq Ch^{\frac{5}{2}}|v|_{1,D}(h^{-1}|\log h|)^{\frac{1}{2}}\left(\int_0^{x_0} r^{4-\sigma}(|\partial_1^2 u(\tau)|^2 + r^2|\partial_1^3 u(\tau)|^2)\,d\tau\right)^{\frac{1}{2}}.$$

By Theorem 3.1.1 we have

$$|I_1| \leq Ch^2|\log h|^{\frac{1}{2}}\|u\|_{4(\sigma),D}|v|_{1,D}.$$

If $\alpha = 2\pi$, we take $0 < \sigma < 1$, then by the corollary of Theorem 3.1.1 we obtain

$$|I_1| \leq Ch^{\frac{5}{2}}|v|_{1,D}\sum_{k=1}^\infty \xi^{\frac{\sigma-1}{2}k}|\log h_k|^{\frac{1}{2}}|\xi^k h\log(\xi^k h)|^{\frac{1}{2}}\|u\|_{4(\sigma),D}$$

$$\leq Ch^2|\log h|\cdot\|u\|_{4(\sigma),D}|v|_{1,D}.$$

In short

$$|A_2| \leq Ch^2|\log h|^\gamma\|u\|_{4(\sigma),D}|v|_{1,D}.$$

Step 4. Estimate for A_3 and the completion of the proof.

The estimate for A_3 follows the same lines as that for A_1, and it is proved that

$$|A_3| \leq Ch^2|\log h|^{\frac{1}{2}}\|u\|_{3(\sigma),D}|v|_{1,D}.$$

Finally we obtain (3.5.11). □

Theorem 3.5.1. *If the conditions* (A),(E) *hold,* $u \in H^{4(\sigma)}(\Omega)$ *and* $u_h \in S(\Omega)$ *are solutions to* (3.3.1) *and* (3.3.2)$_1$ *respectively, and* $u_0^I = \Pi u_0$, *then*

$$\|u_h - \Pi u\|_{1,\Omega} \leq Ch^2|\log h|^\gamma\|u\|_{4(\sigma),\Omega},$$

where the constants σ *and* γ *are the same as those in Lemma* 3.5.5.

Proof. We take $v = u_h - \Pi u$, and by Lemma 3.5.5 we have

$$|u_h - \Pi u|_{1,\Omega}^2 = a(u_h - \Pi u, u_h - \Pi u)$$
$$= a(u - \Pi u, u_h - \Pi u)$$
$$\leq Ch^2|\log h|^\gamma\|u\|_{4(\sigma),\Omega}|u_h - \Pi u|_{1,\Omega}.$$

After cancelling a factor $|u_h - \Pi u|_{1,\Omega}$, then applying the Friedrichs inequality, we reach the conclusion of the theorem. □

Corollary 1. *Under the hypotheses of Theorem 3.5.1, if we assume in addition that* $u \in W^{2,p(\tau)}(\Omega)$, $p > 1$, $2 - p < \tau < 2 - (1 - \frac{\pi}{\alpha})p$, *then*

$$\|u - u_h\|_{0,p,\Omega} \leq C\sqrt{p}h^2|\log h|^\gamma (|u|_{2,p(\tau),\Omega} + \|u\|_{4(\sigma),\Omega}). \tag{3.5.17}$$

Proof. By Theorem 3.2.2 we get

$$\|u - \Pi u\|_{0,p,\Omega} \leq Ch^2|u|_{2,p(\tau),\Omega}. \tag{3.5.18}$$

Then by Theorem 3.1.1 and Theorem 3.5.1 we get

$$\|u - \Pi u\|_{0,p,\Omega} \leq C\sqrt{p}\|u_h - \Pi u\|_{1,\Omega}$$
$$\leq C\sqrt{p}h^2|\log h|^\gamma \|u\|_{4(\sigma),\Omega}. \tag{3.5.19}$$

Finally by (3.5.18),(3.5.19) we get (3.5.17). □

Corollary 2. *Under the hypotheses of Theorem 3.5.1*

$$|(u - u_h)(x)| \leq Ch^2|\log h|^\gamma (|\log r| + |\log h|)^{\frac{1}{2}} \|u\|_{4(\sigma),\Omega}, \quad \forall x \in \Omega. \tag{3.5.20}$$

Proof. Let $e_i \subset T_k$, then by Theorem 3.1.2 we have

$$\|u_h - \Pi u\|_{0,\infty,e_i} \leq C|\log h_k|^{\frac{1}{2}} \|u_h - \Pi u\|_{1,\Omega}$$
$$= C(|\log \xi^k| + |\log h|)^{\frac{1}{2}} \|u_h - \Pi u\|_{1,\Omega}.$$

By Theorem 3.2.4 we get

$$\|u - \Pi u\|_{0,\infty,\Omega} \leq Ch^2|u|_{3(\sigma),\Omega}.$$

Finally applying Theorem 3.5.1 we obtain (3.5.20). □

3.6 Convergence by terms near the singular point

Being the same as the previous section we assume that the partition is similar near the point O, Ω_0 is divided into Ω^* and Ω, and convensional finite element partition is made on Ω^*. We do not require the partition satisfies the condition (E), instead, we give some other conditions.

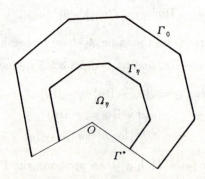

Fig. 34

We introduce some notations first. Divided $\partial\Omega$ into two parts, where the union of two adjacent sides to the point O is denoted by Γ^*, and $\partial\Omega \setminus \Gamma^*$ is denoted by Γ_0 (Fig.34). According to the manner of partition, the similar figures to Γ_0 are $\Gamma_1, \Gamma_2, \cdots, \Gamma_k, \cdots$. Our conditions read as follows:

(F) On the process of refinement Γ_0 keeps fixed, namaly, the domains Ω and Ω^* keep fixed.

(G) There is a constant $\eta \in (0,1)$, such that the parameter ξ of every refinement admits that $\log\eta/\log\xi$ is an integer, namely, the similar figure Γ_η to Γ_0 with centre of similarity O, and constant of proportionality η is always the boundary curve of elements of the partition (Fig. 34).

(H) The angles between Γ_0 and Γ^* are not greater than $\frac{\pi}{2}$.

We still denote by $S(\Omega) \subset H^1(\Omega)$ the infinite element space with linear triangular elements, and define

$$H^1_{\Gamma^*}(\Omega) = \{u \in H^1(\Omega); u|_{\Gamma^*} = 0\},$$

$$S_{\Gamma^*}(\Omega) = S(\Omega) \bigcap H^1_{\Gamma^*}(\Omega).$$

Consider the following auxiliary problem: for given $u_0 \in S_{\Gamma^*}(\Omega)$, find $u \in H^1_{\Gamma^*}(\Omega)$, such that $u|_{\Gamma_0} = u_0|_{\Gamma_0}$, and

$$a(u,v) = 0, \quad \forall v \in H^1_0(\Omega). \tag{3.6.1}$$

The infinite element approximation to (3.6.1) is: find $u_h \in S_{\Gamma^*}(\Omega)$, such that $u_h|_{\Gamma_0} = u_0|_{\Gamma_0}$, and

$$a(u_h, v) = 0, \quad \forall v \in S_0(\Omega). \tag{3.6.2}$$

Lemma 3.6.1. *If the conditions* (A),(F) *hold, and* u, u_h *are the solutions to the problems* (3.6.1) *and* (3.6.2) *respectively, then*

$$\|u - u_h\|_{0,\Omega} \leq Ch|u_0|_{1,\Omega}. \tag{3.6.3}$$

Proof. We apply (3.3.10) of Theorem 3.3.2, where the domain Ω_0 is replaced by the domain Ω, and obtain

$$\|u - u_h\|_{0(-\tilde{\sigma}),\Omega} \leq Ch|u - u_h|_{1,\Omega}. \tag{3.6.4}$$

By (3.6.1) we have

$$a(u,u) = \inf_{\substack{v \in H^1_{\Gamma^*}(\Omega) \\ v|_{\Gamma_0} = u_0|_{\Gamma_0}}} a(u,v),$$

hence

$$|u|_{1,\Omega} \leq |u_0|_{1,\Omega}. \tag{3.6.5}$$

In like manner we have

$$|u_h|_{1,\Omega} \leq |u_0|_{1,\Omega}.$$

Upon substituting them into (3.6.4) and dropping the weight $r^{\tilde{\sigma}}$, we obtain (3.6.3). □

Denote by Ω_η the domain surrounded by Γ_η and Γ^*, then we have

Lemma 3.6.2. *If the conditions* (A),(F),(G) *hold, and* u, u_h *are the solutions to the problems* (3.6.1) *and* (3.6.2) *respectively, then*

$$\|u - u_h\|_{1,\Omega_\eta} \leq Ch|u_0|_{1,\Omega}. \tag{3.6.6}$$

Proof. By (3.6.5) and the Friedrichs inequality we get

$$\|u\|_{1,\Omega} \leq C|u_0|_{1,\Omega}. \tag{3.6.7}$$

$\|u_h\|_{1,\Omega}$ possesses the same upper bound, therefore it will suffice to prove (3.6.6) for sufficiently small h.

Take constants $\eta_j, j = 1,2,3,4$, such that $0 < \eta_1 < \eta_2 < \eta < \eta_3 < \eta_4 < 1$. We draw similar figures of Γ_0 with centre of similarity O and the constants of proportionality η_j, denoted by Γ_{η_j}. Denote by Ω_{η_j} the domain surrounded by Γ_{η_j} and Γ^*, by $\Omega(\eta_1, \eta)$ a subdomain of Ω lies between Γ_{η_1} and Γ_η, and so on. For the sake of simplicity and without losing generality we assume that Γ_{η_2} is the boundary curve of elements of the partition.

Applying interior estimate [46] we obtain

$$\|u\|_{2,\Omega(\eta_1,\eta_4)} \leq C\|u\|_{1,\Omega}. \tag{3.6.8}$$

The partition is finite on $\Omega(\eta_1, \eta_4)$, so we can apply the interior estimate for the finite element method [59] and obtain

$$\|u - u_h\|_{1,\Omega(\eta_2,\eta_3)} \leq C(h\|u\|_{2,\Omega(\eta_1,\eta_4)} + \|u - u_h\|_{0,\Omega(\eta_1,\eta_4)}). \tag{3.6.9}$$

We define an auxiliary function $\zeta \in C^\infty(\bar{\Omega}_{\eta_3})$ with the following properties: $0 \leq \zeta(x) \leq 1$, $\zeta \equiv 0$, on the domain Ω_{η_2}, and $\zeta \equiv 1$ on the domain $\Omega(\eta, \eta_3)$. Let

$$v^I = \Pi(\zeta(u - u_h)),$$

then $v^I|_{\Gamma_\eta} = \Pi u - u_h$. Let us estimate $|v^I|_{1,\Omega_\eta}$. We take an arbitrary element $e_i \subset \Omega_{\eta_3}$. Let b_1, b_2, b_3 be three nodes on it. Denote by $\lambda_1, \lambda_2, \lambda_3$ the corresponding barycentric coordinates. Let

$$w_1 = \zeta(b_1) \sum_{j=1}^{3} (u(b_j) - u_h(b_j))\lambda_j,$$

$$w_2 = \sum_{j=2}^{3} (\zeta(b_j) - \zeta(b_1))(u(b_j) - u_h(b_j))\lambda_j,$$

then $v^I|_{e_i} = w_1 + w_2$. We have

$$|w_1|_{1,e_i} = \zeta(b_1)|\Pi u - u_h|_{1,e_i} \leq |\Pi u - u_h|_{1,e_i}, \tag{3.6.10}$$

$$|w_2|_{1,e_i} \leq \sum_{j=2}^{3} |\zeta(b_j) - \zeta(b_1)| \cdot |u(b_j) - u_h(b_j)| \cdot |\lambda_j|_{1,e_i}.$$

$|\lambda_j|_{1,e_i}$ is bounded, and by the smoothness of ζ,

$$|\zeta(b_j) - \zeta(b_1)| < Ch,$$

and by Lemma 3.4.3 we have

$$|u(b_j) - u_h(b_j)| \leq Ch^{-1} \|\Pi u - u_h\|_{0,e_i},$$

therefore

$$|w_2|_{1,e_i} \leq C \|\Pi u - u_h\|_{0,e_i}. \qquad (3.6.11)$$

If h is sufficiently small, then the set of e_i covers the domain Ω_η. Besides we notice that $v^I|_{\Omega_{\eta_2}} = 0$, then by (3.6.10),(3.6.11) we get

$$|v^I|_{1,\Omega_\eta} \leq C \|\Pi u - u_h\|_{1,\Omega(\eta_2,\eta_3)}.$$

By the Friedrichs inequality we have

$$\|v^I\|_{1,\Omega_\eta} \leq C \|\Pi u - u_h\|_{1,\Omega(\eta_2,\eta_3)}. \qquad (3.6.12)$$

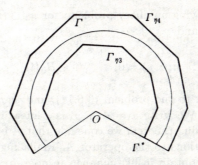

Fig 35

We costruct a curve Γ in the domain $\Omega(\eta_3, \eta_4)$, such that Γ and Γ^* surround a domain Ω_Γ, the boundary of which belongs to C^∞ everywhere except tho point O, cf. Fig.35. Applying (3.6.7),(3.6.8) and the trace theorem we obtain

$$\|u\|_{\frac{3}{2},\Gamma} \leq C |u_0|_{1,\Omega}.$$

On the analogy of Lemma 3.3.1, we can prove

$$\|u\|_{2(\sigma),\Omega_\Gamma} \leq C \|u\|_{\frac{3}{2},\Gamma},$$

hence
$$\|u\|_{2(\sigma),\Omega_\Gamma} \leq C|u_0|_{1,\Omega}. \qquad (3.6.13)$$

By Theorem 3.3.1 and (3.6.12) we have
$$\|u - u_h\|_{1,\Omega_\eta} \leq C(h|u|_{2(\sigma),\Omega_\eta} + \|\Pi u - u_h\|_{1,\Omega(\eta_2,\eta_3)}).$$

Then using the triangle inequality we get
$$\|\Pi u - u_h\|_{1,\Omega(\eta_2,\eta_3)} \leq \|\Pi u - u\|_{1,\Omega(\eta_2,\eta_3)} + \|u - u_h\|_{1,\Omega(\eta_2,\eta_3)}.$$

By Theorem 3.2.2 we have
$$\|\Pi u - u\|_{1,\Omega(\eta_2,\eta_3)} \leq Ch|u|_{2(\sigma),\Omega_\Gamma}$$

for sufficiently small h. Consequently
$$\|u - u_h\|_{1,\Omega_\eta} \leq C(h|u|_{2(\sigma),\Omega_\Gamma} + \|u - u_h\|_{1,\Omega(\eta_2,\eta_3)}).$$

Upon substituting (3.6.13),(3.6.9),(3.6.8),(3.6.7) into it we obtain
$$\|u - u_h\|_{1,\Omega_\eta} \leq C(h|u_0|_{1,\Omega} + \|u - u_h\|_{0,\Omega(\eta_1,\eta_4)}).$$

Finally substituting (3.6.3) into it yields (3.6.6). □

We consider the following auxiliary problem: for $u_0 \in H^1_{\Gamma^*}(\Omega)$, find $u \in H^1_{\Gamma^*}(\Omega)$, such that $u|_{\Gamma_0} = u_0|_{\Gamma_0}$, and
$$a(u,v) = 0, \qquad \forall v \in H^1_0(\Omega). \qquad (3.6.14)$$

Let u be the solution to the problem (3.6.14), and $w(x) = u(\eta x), \forall x \in \Omega$. It is clear that $w \in H^1_{\Gamma^*}(\Omega)$. We write $w = Au_0$. A is a linear bounded operator. Using (3.6.13) and the embedding theorem we conclude that A is a compact operator.

We define approximations to the operator A. Replacing the problem (3.6.14) by the problem (3.6.2), we obtain in like manner an operator $A_h : S_{\Gamma^*}(\Omega) \to S_{\Gamma^*}(\Omega)$. Since u_h is a solution to the infinite element method in (3.6.2), using the approach in Chapter One, we can express the values at nodes as
$$y_k = X^k y_0,$$

where X is the transfer matrix. Therefore the range of definition of A_h is a finite dimentional space.

We extend A_h to the space $H^1_{\Gamma^*}(\Omega)$ in two different ways. We define an orthogonal projection operator $P : H^1_{\Gamma^*}(\Omega) \to S_{\Gamma^*}(\Omega)$ with the inner product $a(\cdot,\cdot)$ in the space $H^1_{\Gamma^*}(\Omega)$. Let $A_h = A_h P + A(I - P)$ and $A_h^* = A_h P$ on $H^1_{\Gamma^*}(\Omega)$.

Lemma 3.6.3. *If the conditions* (A),(F),(G),(H) *hold, then*

$$\|A - A_h\| \le Ch, \qquad (3.6.15)$$

$$\|A - A_h^*\| \le Ch^{\frac{1}{2}-s}, \qquad (3.6.16)$$

where $0 < s < \frac{1}{2}$, *and the constant* C *in* (3.6.16) *depends on* s.

Proof. We take an arbitrary $u_0 \in H^1_{\Gamma^*}(\Omega)$, and using Lemma 3.6.2 we get

$$\|Au_0 - A_h u_0\|_{1,\Omega} = \|A_h P u_0 - A P u_0\|_{1,\Omega}$$

$$\le Ch|Pu_0|_{1,\Omega} \le Ch|u_0|_{1,\Omega}. \qquad (3.6.17)$$

Thus (3.6.15) is verified. We turn now to prove (3.6.16). By the triangle inequality we have

$$\|Au_0 - A_h^* u_0\|_{1,\Omega} \le \|A(u_0 - Pu_0)\|_{1,\Omega} + \|(A - A_h)Pu_0\|_{1,\Omega}. \qquad (3.6.18)$$

We solve the Laplace equation on Ω with boundary data $(u_0 - Pu_0)|_{\partial\Omega}$, and denote by w the weak solution. Applying interior estimate we get

$$\|w\|_{1,\Omega(\eta_1,\eta_4)} \le C\|w\|_{\frac{1}{2}+s,\Omega}. \qquad (3.6.19)$$

Let $w' = w - (u_0 - Pu_0)$, then $w' \in H^1_0(\Omega)$, which satisfies the equation

$$a(w', v) = -a(u_0 - Pu_0, v), \qquad \forall v \in H^1_0(\Omega). \qquad (3.6.20)$$

Take an arbitrary $\varphi \in C_0^\infty(\Omega)$, and solve the following boundary value problem:

$$\begin{cases} -\triangle v = \varphi, \\ v|_{\partial\Omega} = 0, \end{cases}$$

then v is a classical solution. Denote by $(\cdot,\cdot)$ the L^2-inner product, and use the Green's formula to get

$$(w', \varphi) = (w', -\triangle v) = a(w', v).$$

By (3.6.20) we obtain

$$(w', \varphi) = -\int_\Omega \nabla(u_0 - Pu_0) \cdot \nabla v \, dx.$$

It is easy to prove that

$$\|v\|_{\frac{3}{2}-s,\Omega} \le C\|\varphi\|_{-\frac{1}{2}-s,\Omega},$$

consequently

$$\|\nabla v\|_{\frac{1}{2}-s,\Omega} \le C\|\varphi\|_{-\frac{1}{2}-s,\Omega}.$$

On the other hand
$$\|\nabla(u_0 - Pu_0)\|_{s-\frac{1}{2},\Omega} \leq C\|u_0 - Pu_0\|_{\frac{1}{2}+s,\Omega},$$
therefore
$$|(w',\varphi)| \leq C\|u_0 - Pu_0\|_{\frac{1}{2}+s,\Omega}\|\varphi\|_{-\frac{1}{2}-s,\Omega}, \quad \forall \varphi \in C_0^\infty(\Omega).$$
It is known that $w' \in H_0^1(\Omega)$, hence
$$\|w'\|_{\frac{1}{2}+s,\Omega} \leq C\|u_0 - Pu_0\|_{\frac{1}{2}+s,\Omega}.$$
Applying the triangle inequality we get
$$\|w\|_{\frac{1}{2}+s,\Omega} \leq C\|u_0 - Pu_0\|_{\frac{1}{2}+s,\Omega}. \tag{3.6.21}$$

Following the same lines as the proof of Theorem 3.3.2, we estimate $\|u_0 - Pu_0\|_{0,\Omega}$. We note that
$$a(Pu_0, v) = a(u_0, v), \quad \forall v \in S_{\Gamma^*}(\Omega).$$
Setting $g = r^{\tilde{\sigma}}(u - u_h)$, we consider a mixed boundary value problem
$$\begin{cases} -\triangle \varphi = g, \\ \varphi|_{\Gamma^*} = 0, \quad \frac{\partial \varphi}{\partial \nu}|_{\Gamma_0} = 0. \end{cases}$$
On the analogy of Lemma 3.3.1 we can prove
$$\|\varphi\|_{2(\sigma),\Omega} \leq C\|g\|_{0(\tilde{\sigma}),\Omega}.$$
Now the second term of the right hand side of (3.3.9) vanishes, that is
$$\|u - u_h\|_{0(-\tilde{\sigma}),\Omega}^2 = a(u - u_h, \varphi - \Pi\varphi).$$
On the analogy of (3.3.10) we can prove
$$\|u_0 - Pu_0\|_{0(-\tilde{\sigma}),\Omega} \leq Ch|u_0 - Pu_0|_{1,\Omega}.$$
We drop the weight function $r^{\tilde{\sigma}}$ and notice that
$$|u_0 - Pu_0|_{1,\Omega} \leq |u_0|_{1,\Omega},$$
then we have
$$\|u_0 - Pu_0\|_{0,\Omega} \leq Ch|u_0|_{1,\Omega}.$$
We also have
$$\|u_0 - Pu_0\|_{1,\Omega} \leq C|u_0|_{1,\Omega}.$$

Applying the interpolation inequality [55] we get

$$\|u_0 - Pu_0\|_{\frac{1}{2}+s,\Omega} \le Ch^{\frac{1}{2}-s}|u_0|_{1,\Omega}.$$

Substituting it into (3.6.21) we obtain

$$\|w\|_{\frac{1}{2}+s,\Omega} \le Ch^{\frac{1}{2}-s}|u_0|_{1,\Omega}.$$

Then we substitute it into (3.6.19) and obtain

$$\|w\|_{1,\Omega(\eta_1,\eta_4)} \le Ch^{\frac{1}{2}-s}|u_0|_{1,\Omega}.$$

The trace theorem leads to

$$\|w\|_{\frac{1}{2},\partial\Omega_\eta} \le Ch^{\frac{1}{2}-s}|u_0|_{1,\Omega}.$$

w is the weak solution to the Laplace equation on Ω_η, hence

$$\|w\|_{1,\Omega_\eta} \le Ch^{\frac{1}{2}-s}|u_0|_{1,\Omega},$$

consequently

$$\|A(u_0 - Pu_0)\|_{1,\Omega} \le Ch^{\frac{1}{2}-s}|u_0|_{1,\Omega}.$$

By (3.6.17),(3.6.18) we get

$$\|Au_0 - A_h^*u_0\|_{1,\Omega} \le Ch^{\frac{1}{2}-s}|u_0|_{1,\Omega},$$

which proves (3.6.16). □

We evaluate the eigenvalues and eigenfunctions of the operator A according to the expression (3.0.2). Let λ and u_0 be the eigenvalue and the eigenfunction of the operator A, $\lambda \ne 0$, then

$$\lambda u_0 = Au_0. \tag{3.6.22}$$

According to the definition Au_0 is the solution on Ω to the Laplace equation and the boundary condition $Au_0|_{\Gamma^*} = 0$. (3.6.22) implies u_0 is also a solution. Then the equation (3.6.14) implies $u = u_0$. By (3.6.22) we have

$$\lambda u_0(x) = u_0(\eta x), \quad \forall x \in \Omega. \tag{3.6.23}$$

u_0 admits the expansion (3.0.2) near the point O. Substituting (3.0.2) into (3.6.23) we get

$$\lambda \sum_{j=1}^{\infty} c_j r^{\frac{j\pi}{\alpha}} \sin\frac{j\pi}{\alpha}(\vartheta - \vartheta_0) = \sum_{j=1}^{\infty} c_j (\eta r)^{\frac{j\pi}{\alpha}} \sin\frac{j\pi}{\alpha}(\vartheta - \vartheta_0).$$

Comparing the coefficients on the both sides, we see that there must be $j \in \mathbb{N}$ and a constant $c_j \neq 0$ such that

$$\lambda = \eta^{\frac{j\pi}{\alpha}}, \quad u_0 = c_j r^{\frac{j\pi}{\alpha}} \sin \frac{j\pi}{\alpha}(\vartheta - \vartheta_0). \tag{3.6.24}$$

Conversely it is easy to see that λ and u_0 given by (3.6.24) are the eigenvalue and eigenfunction of the operator A. They are denoted by λ_j and u_j respectively in the following.

Let $\eta = \xi^k$. Denote by X the transfer matrix defined in Chapter One. Let $\lambda_{h\,1}, \lambda_{h\,2}, \ldots$ be eigenvalues of the matrix X^k arranged in an order from large norm to small. A vector y_0 can be represented as a linear combination of eigenvectors of X^k, thus the infinite element solution admits an expansion

$$u_h = \sum_j u_{h\,j}, \tag{3.6.25}$$

where $u_{h\,j}$ corresponds to $\lambda_{h\,j}$.

Theorem 3.6.1. *We assume the conditions (A),(F),(G),(H) hold. Fix a $j \in \mathbb{N}$, then there are constants C and $h_0 > 0$, such that*

$$|\lambda_j - \lambda_{h\,j}| \leq Ch \tag{3.6.26}$$

for $h \leq h_0$. Moreover if u and u_h are the solutions to the problems (3.3.1),(3.3.2)$_1$ respectively, and (3.0.2),(3.6.25) are the expansions of them near the point O, then

$$\|u_j - u_{h\,j}\|_{1,\Omega} \leq C\{\|u - u_h\|_{1,\Omega} + h\|u\|_{1,\Omega}\} \tag{3.6.27}$$

for $h \leq h_0$.

Proof. Denote by $R(z, A), R(z, A_h), R(z, A_h^*)$ the resolvents of the operators A, A_h, A_h^* on the complex plane, then the poles of $R(z, A)$ are $\lambda_j, j = 1, 2, \cdots$. We construct a circle $\mathcal{C}$, such that $\lambda_{j+1}, \lambda_{j+2}, \cdots$ and $\lambda = 0$ are situated at the interior of $\mathcal{C}$, and $\lambda_1, \cdots, \lambda_j$ the exterior of $\mathcal{C}$. By the perturbation theory for operators [49], and Lemma 3.6.3, there is a constant $h_1 > 0$, such that the resolvents $R(z, A_h)$, $R(z, A_h^*)$ are always analytic near $\mathcal{C}$, and each of A_h and A_h^* has exact j eigenvalues outside $\mathcal{C}$, for $h \leq h_1$. Since the eigenvalues of A_h^* are also the eigenvalues of A_h, the eigenvalues of A_h^* and A_h coincide outside $\mathcal{C}$.

It is easy to see that the eigenvalues of A_h^* are the same as those of X^k, except $\lambda = 0$. Exactly they are $\lambda_{h\,1}, \ldots, \lambda_{h\,j}$ outside $\mathcal{C}$.

We construct circles $\mathcal{C}_j$ with centre λ_j and radius δ, $0 < \delta < |\lambda_j - \lambda_{j+1}|$, then λ_j is the unique eigenvalue of A in this circle. $z = \lambda_j$ is a pole of $R(z, A)$ with order one, hence on $\mathcal{C}_j$ we have

$$\|R(z, A)\| \leq \frac{C_1}{\delta}.$$

3.6 CONVERGENCE BY TERMS

By Lemma 3.6.3 we get
$$\|A - A_h\| \le C_2 h.$$
Let $h_0 = \min\{h_1, \frac{|\lambda_j - \lambda_{j+1}|}{C_1 C_2 + 2}\}$, and take $\delta = (C_1 C_2 + 1)h$, then
$$\|A - A_h\| \le \frac{\delta}{C_1}$$
provided $h \le h_0$. Therefore λ_{hj} is at the interior of $\mathcal{C}_j$, that is
$$|\lambda_j - \lambda_{hj}| < \delta = (C_1 C_2 + 1)h,$$
which proves (3.6.26).

Now we fix a $\delta, 0 < \delta < |\lambda_j - \lambda_{j+1}|$, and let
$$P_j(A) = -\frac{1}{2\pi i} \int_{\mathcal{C}_j} R(z, A)\, dz,$$
$$P_j(A_h) = -\frac{1}{2\pi i} \int_{\mathcal{C}_j} R(z, A_h)\, dz,$$
$$P_j(A_h^*) = -\frac{1}{2\pi i} \int_{\mathcal{C}_j} R(z, A_h^*)\, dz,$$
then they are projections from $H^1_{\Gamma^*}(\Omega)$ to the eigenspaces [49]. These eigenspaces are isomorphic to each other for $h \le h_0$, hence all of them are one dimensional. From the expansions (3.0.2),(3.6.25) we get
$$u_j = P_j(A)u, \qquad u_{hj} = P_j(A_h^*)u_h.$$
But $u_h \in S_{\Gamma^*}(\Omega)$ and $A_h^* = A_h$ on $S_{\Gamma^*}(\Omega)$, therefore we also get
$$u_{hj} = P_j(A_h)u_h.$$
By Lemma 3.6.3 and the boundedness of the operator $P_j(A_h)$ we get
$$\|P_j(A)u - P_j(A_h)u_h\|_{1,\Omega}$$
$$\le \|P_j(A)u - P_j(A_h)u\|_{1,\Omega} + \|P_j(A_h)u - P_j(A_h)u_h\|_{1,\Omega}$$
$$\le C\{h\|u\|_{1,\Omega} + \|u - u_h\|_{1,\Omega}\},$$
which is (3.6.27). □

We have obtained the estimate for $\|u - u_h\|_{1,\Omega_0}$ in §3.3, then combining it with Theorem 3.6.1 we obtain the term by term error estimate for the expansions. Theorem 3.6.1 also implies estimation for exponents. It is easy to see that the j-th term of the infinite element solution, $u_{hj} \sim r^{\alpha_j}$, where
$$\alpha_j = \frac{\log \lambda_{hj}}{\log \eta},$$

then (3.6.26) gives
$$\left|\alpha_j - \frac{j\pi}{\alpha}\right| \leq Ch.$$

We have given a method to calculate the stress intensity factors in §1.9. Concerning to the problem (3.0.1) it is: We intend to evaluate the approximate values to c_j of (3.0.2), and set

$$A_j = a\left(r^{\frac{j\pi}{\alpha}}\sin\frac{j\pi}{\alpha}(\vartheta-\vartheta_0), r^{\frac{j\pi}{\alpha}}\sin\frac{j\pi}{\alpha}(\vartheta-\vartheta_0)\right),$$

then
$$a(u_j, u_j) = c_j^2 A_j.$$

On the other hand if the infinite element solution is represented as a linear combination of the eigenvectors $g_1, \ldots, g_n$ as follows,

$$y_0 = \sum_{j=1}^{n} d_j g_j,$$

then using the notations of combined stiffness mstrix in Chapter One we have

$$a(u_{h\,j}, u_{h\,j}) = d_j^2 g_j^T K_z g_j.$$

Thus we can define

$$c_{h\,j}^2 = \frac{1}{A_j} d_j^2 g_j^T K_z g_j,$$

which are the approximate values to c_j^2. The estimate

$$|c_j^2 - c_{h\,j}^2| = \frac{1}{A_j}|a(u_j, u_j) - a(u_{h\,j}, u_{h\,j})|$$
$$= \frac{1}{A_j}|a(u_j, u_j) - a(u_j, u_{h\,j}) + a(u_j, u_{h\,j}) - a(u_{h\,j}, u_{h\,j})|$$
$$\leq \frac{1}{A_j}(|u_j|_{1,\Omega} + |u_{h\,j}|_{1,\Omega})|u_j - u_{h\,j}|_{1,\Omega}$$

holds. Applying the Corollary of Theorem 3.3.1 and Theorem 3.6.1 we obtain

$$|c_j^2 - c_{h\,j}^2| = O(h).$$

3.7 Multigrid algorithm (II)

We consider the same problem and mesh as those in §3.6. The set of triangular elements is denoted by T_1, which is the coarse mesh.

As we have stated in §1.13, let $\mathcal{T}_2, \cdots, \mathcal{T}_J$ be fine meshes. $\mathcal{T}_{j+1}$ is a refinement of $\mathcal{T}_j, j = 1, 2, \cdots, J - 1$. The constants of proportionality for $\mathcal{T}_j$ are $\xi^{\frac{1}{2^j}}, \xi^{\frac{2}{2^j}}, \cdots$, $\xi^{\frac{k}{2^j}}, \cdots$. On the domain Ω^* from $\mathcal{T}_j$ to $\mathcal{T}_{j+1}$ each element is divided into four triangles appropriately.

For simplicity we assume that the function f in (3.0.1) is piecewise linear in accordance with the mesh $\mathcal{T}_1$. We define a function u_0 on $\bar{\Omega}_0$ such that it is continuous on $\bar{\Omega}_0$, $u_0|_{\partial \Omega_0} = f$, and it is linear on all $e_i \in \mathcal{T}_1$ and vanishes at all interior nodes. Let $u - u_0$ be a new variable, still denoted by u, then it is the solution to

$$\begin{cases} -\triangle u = g, & x \in \Omega_0, \\ u|_{\partial \Omega_0} = 0, \end{cases} \qquad (3.7.1)$$

where g vanishes near the singular point O.

We obtain nested subspaces of $H_0^1(\Omega_0)$, $M_1(\Omega_0) \subset M_2(\Omega_0) \subset \cdots \subset M_J(\Omega_0)$, associated with $\mathcal{T}_1, \ldots, \mathcal{T}_J$. The infinite element problems are: find $u_j \in M_j(\Omega_0)$, such that

$$a(u_j, v) = (g, v), \qquad \forall v \in M_j(\Omega_0). \qquad (3.7.2)$$

In accordance with the partition on Ω^*, we have finite dimensional subspaces of $\{w \in H^1(\Omega^*); w|_{\partial \Omega^* \cap \partial \Omega_0} = 0\}$, $M_1(\Omega^*) \subset M_2(\Omega^*) \subset \cdots \subset M_J(\Omega^*)$. Let K_z^j be the combined stiffness matrix for $j, 1 \leq j \leq J$. By Lemma 2.9.1, the problem (3.7.2) is equivalent to: find $u_j \in M_j(\Omega^*)$, such that

$$a(u, v)_{\Omega^*} + y_0^T K_z^j z_0 = (g, v)_{\Omega^*}, \qquad \forall v \in M_j(\Omega^*), \qquad (3.7.3)$$

where $y_0 = B_0 u, z_0 = B_0 v$.

We define inner product

$$(u, v)_j = h_j^2 \sum_{i=1}^{n_j} u(x_i^j) v(x_i^j)$$

on the space $M_j(\Omega^*)$, where h_j is the maximum length of the sides of triangular elements on $\Omega^* \cup \Omega_1$, x_i^j are nodes on Ω^*, and n_j is the number of nodes. Let $\|u\|_j^2 = (u, u)_j$. We assume that the meshes satisfy, besides the conditions (A) and (B), the condition,

(I) The meshes on Ω^* are quasi-uniform, that is, the ratio of the longest side to the shortest side of $\mathcal{T}_j$ is bounded by a constant independent of j.

Under these conditions the norm $\|\cdot\|_j$ is equivalent to the L^2-norm. According to (3.7.3), we define a bilinear form,

$$a_j(u, v) = a(u, v)_{\Omega^*} + y_0^T K_z^j z_0. \qquad (3.7.4)$$

We need to define a prolongation operator $I_j : M_{j-1}(\Omega^*) \to M_j(\Omega^*)$. Here we use a typical prolongation, which is $I_j v \equiv v$ for all $v \in M_{j-1}(\Omega^*)$. Then we define two restriction operators, $I_j^t : M_j(\Omega^*) \to M_{j-1}(\Omega^*)$, $I_j^* : M_j(\Omega^*) \to M_{j-1}(\Omega^*)$ as follows:

$$(I_j^t w, \phi)_{j-1} = (w, I_j \phi)_j, \quad \forall w \in M_j(\Omega^*), \quad \phi \in M_{j-1}(\Omega^*), \tag{3.7.5}$$

$$a_{j-1}(I_j^* w, \phi) = a_j(w, I_j \phi), \quad \forall w \in M_j(\Omega^*), \quad \phi \in M_{j-1}(\Omega^*). \tag{3.7.6}$$

The operator $A_j : M_j(\Omega^*) \to M_j(\Omega^*)$ is defined as:

$$(A_j w, \phi)_j = a_j(w, \phi), \forall \phi \in M_j(\Omega^*).$$

Lemma 3.7.1. *If* $\lambda_{\max}, \lambda_{\min}$ *are the maximum and minimum eigenvalues of* A_j, *then*

$$\lambda_{\max} \leq C h_j^{-2}, \qquad \lambda_{\min} \geq C^{-1}.$$

Proof. It will suffice to estimate

$$\frac{a_j(u,u)}{(u,u)_j}, \qquad \forall u \in M_j(\Omega^*).$$

For a given $u \in M_j(\Omega^*)$ we extend it to Ω as we did in the proof of Lemma 2.9.1, then we get a function in $M_j(\Omega_0)$, still denoted by u. By (3.7.4) we have

$$a_j(u,u) = a(u,u)_{\Omega_0}. \tag{3.7.7}$$

Let us extend u in another way. Let $v \equiv 0$ on $\xi\Omega$, $v \equiv u$ on Ω^*, and $v \in M_j(\Omega_0)$. Since u makes the functional reach its minimum value, we have

$$a(u,u)_\Omega \leq a(v,v)_\Omega. \tag{3.7.8}$$

We investigate v on Ω_1. Let $e \subset \Omega_1$ be a triangular element, then e possesses either a common side s with Γ_0, or a common node x. For the former case we construct an affine mapping such that e is mapped to a reference element $\hat{e}$. Let the image of s be $\hat{s}$ under this mapping. Since $v = 0$ at one node, we have

$$|v|_{1,\hat{e}}^2 \leq C \|v\|_{0,\hat{s}}^2.$$

Applying the properties of affine mapping [61], we have

$$|v|_{1,e}^2 \leq C |v|_{1,\hat{e}}^2, \quad \|v\|_{0,\hat{s}}^2 \leq C h_j^{-1} \|v\|_{0,s}^2,$$

thus

$$|v|_{1,e}^2 \leq C h_j^{-1} \|v\|_{0,s}^2. \tag{3.7.9}$$

For the later case by the same reason we have

$$|v|_{1,e}^2 \leq C v^2(x).$$

Suppose $x \in s \subset \Gamma_0$, then because the interpolating function is linear, we get

$$v^2(x) \leq Ch_j^{-1}\|v\|_{0,s}^2.$$

(3.7.9) also follows. By addition with respect to elements, we get

$$|v|_{1,\Omega_1}^2 \leq Ch_j^{-1}\|v\|_{0,\Gamma_0}^2,$$

that is

$$a(v,v)_\Omega \leq Ch_j^{-1}\|v\|_{0,\Gamma_0}^2.$$

Applying (3.7.8) we have

$$a(u,u)_\Omega \leq Ch_j^{-1}\|u\|_{0,\Gamma_0}^2. \tag{3.7.10}$$

Next let us investigate the elements e neighboring Γ_0 in Ω^*. Let the common side be s, then using affine mapping we obtain

$$\|u\|_{0,s}^2 \leq Ch_j^{-1}\|u\|_{0,e}^2$$

in like manner. Then upon addition we obtain

$$\|u\|_{0,\Gamma_0}^2 \leq Ch_j^{-1}\|u\|_{0,\Omega^*}^2.$$

In conjunction with (3.7.10) it gives

$$a(u,u)_\Omega \leq Ch_j^{-2}\|u\|_{0,\Omega^*}^2.$$

We have the inverse inequality [61],

$$a(u,u)_{\Omega^*} \leq Ch_j^{-2}\|u\|_{0,\Omega^*}^2,$$

therefore

$$a_j(u,u) \leq Ch_j^{-2}\|u\|_{0,\Omega^*}^2.$$

Notig that $\|\cdot\|_{0,\Omega^*}$ is equivalent to $\|\cdot\|_j$, we get the estimate for $\lambda_{\max}$.

By the Poincare inequality

$$\|u\|_{0,\Omega^*}^2 \leq Ca(u,u)_{\Omega^*} \leq Ca_j(u,u),$$

which gives the estimate for $\lambda_{\min}$. □

We will make a basic assumption on the smoother R_j introduced in §1.13. Let $K_j = I - R_j A_j$, and K_j^* be the adjoint operator of K_j with respect to the inner product $(A_j \cdot, \cdot)$. We assume that

$$\|A_j v\|_j^2 \leq C\lambda_j a_j((I - K_j^* K_j)v, v), \quad \forall v \in M_j(\Omega^*), \tag{3.7.11}$$

where λ_j is the maximum eigenvalue of A_j. Applying Lemma 3.7.1 one can verify that the inequality (3.7.11) holds for Richardson iteration and Gauss-Seidel iteration [67].

Lemma 3.7.2.

$$|a_j((I - I_j I_j^*)u, u)| \leq C\lambda_j^{-\frac{1}{2}} \|A_j u\|_j a_j(u,u)^{\frac{1}{2}}, \quad \forall u \in M_j(\Omega^*), \qquad (3.7.12)$$

where λ_j is the maximum eigenvalue of A_j.

Proof.

$$\begin{aligned} |a_j((I-I_j I_j^*)u, u)| &= |((I - I_j I_j^*)u, A_j u)_j| \\ &\leq \|(I - I_j I_j^*)u\|_j \|A_j u\|_j \\ &\leq C\|u - I_j^* u\|_{0,\Omega^*} \|A_j u\|_j. \end{aligned} \qquad (3.7.13)$$

We extend u to Ω as we did before. Then (3.7.6) implies that $I_j^* u$ is the infinite element solution on the mesh $\mathcal{T}_{j-1}$. Applying the estimate (3.3.10) we obtain

$$\begin{aligned} \|u - I_j^* u\|_{0,\Omega^*} &\leq Ch_{j-1} |u - I_j^* u|_{1,\Omega_0} \\ &\leq Ch_{j-1} |u|_{1,\Omega_0}. \end{aligned}$$

From (3.7.7) we have

$$|u|_{1,\Omega_0} = a(u,u)_{\Omega_0}^{\frac{1}{2}} = a_j(u,u)^{\frac{1}{2}}.$$

Lemma 3.7.1 implies

$$h_{j-1} = 2h_j \leq C\lambda_j^{-\frac{1}{2}}.$$

Substituting them into (3.7.13), we obtain (3.7.4). □

Lemma 3.7.3.

$$a_j(I_j u, I_j u) \leq a_{j-1}(u, u), \quad \forall u \in M_{j-1}(\Omega^*). \qquad (3.7.14)$$

Proof. We notice (3.7.4). Let y_0 correspond to the mesh $\mathcal{T}_j$, and $\tilde{y}_0$ correspond to $\mathcal{T}_{j-1}$. We only need to verify

$$y_0^T K_z^j y_0 \leq \tilde{y}_0^T K_z^{j-1} \tilde{y}_0.$$

Let u_j and u_{j-1} be the solutions on Ω associated with y_0 and $\tilde{y}_0$ respectively, then it will suffice to verify

$$a(u_j, u_j)_\Omega \leq a(u_{j-1}, u_{j-1})_\Omega.$$

This inequality is obvious, because $a(\cdot,\cdot)_\Omega$ will decrease along with the refinement of the mesh if boundary value keeps fixed. □

The following theorem is quoted from [67] which is a generalization of one theorem in [66].

Theorem 3.7.1. If (3.7.11),(3.7.12) and (3.7.14) hold, then

$$0 \leq a_j(E_j u, u) \leq \delta_j a_j(u,u), \qquad \forall u \in M_j(\Omega^*),$$

where $j = 1, \ldots, J, E_j = I - B_j A_j$, for the V-circle we have

$$\delta_j = \frac{jM}{m^{\frac{1}{2}} + jM};$$

for the generalized V-circle we have

$$\delta_j = \frac{M}{m_j^{\frac{1}{2}} + M};$$

and for the W-circle we have

$$\delta_j = \frac{M^{\frac{1}{2}}}{(m+M)^{\frac{1}{2}}},$$

where $M > 0$ is independent of j.

The above theorem leads to a conclusion: the above multigrid algorithm converges for any times of smoothing iterations, and the contraction number δ_j is independent of j for the W-circle and the generalized V-circle.

3.8 Parabolic equations (II)

In this section the domain and mesh are the same as those in §3.6, but the problem is

$$\frac{\partial u}{\partial t} - \triangle u = f(x), \quad (x,t) \in \Omega \times (0,T], \tag{3.8.1}$$

$$u\big|_{x \in \partial\Omega} = 0, \tag{3.8.2}$$

$$u\big|_{t=0} = u_0(x), \tag{3.8.3}$$

where $f \in L^2(\Omega_0), u_0 \in H^2(\Omega_0)$.

We make one more finite element partition in the domain Ω as we did in §1.16. As usual we require that the two sets of elements on Ω conform to each other, that is, the nodes of them coincide on the curve Γ_0.

We construct two finite dimensional subspaces V an W of $\{w \in H^1(\Omega); w\big|_{\partial\Omega \cap \partial\Omega_0}$ $= 0\}$. Here V is associated with the infinite element partition, given in §3.6, and moreover each function $v \in V$ is already an approximate solution to the Laplace equation, that is, $y_k = X^k y_0, k = 1, \ldots,$ where X is the transfer matrix. W is a convensional finite element subspace associated with finite element partition. A finite element subspace U^* is also constructed for the domain Ω^*. Finally we construct a finite dimensional subspace $S(\Omega_0)$ of $H_0^1(\Omega_0)$, such that if $u \in S(\Omega_0)$, then the restriction of u on Ω^* belongs to U^*, and $u = v + w$ on Ω, where $v \in V$, $w \in W$.

Let Δt be the length of time steps. The infinite element-implicit Euler's scheme to solve (3.8.1)–(3.8.3) is as follows: find $u_h^{(m+1)} \in S(\Omega_0)$ for $m = 0, 1, \ldots$, such that

$$\left(\frac{u_h^{(m+1)} - u_h^{(m)}}{\Delta t}, v\right) + a(u_h^{(m+1)}, v) = (f, v), \quad \forall v \in S_0(\Omega_0), \tag{3.8.4}$$

where $u_h^{(0)}$ is an approximate function to the initial data u_0. We may take $u_h^{(0)} = \Pi u_0$, where Π is the interpolating operator with respect to the finite element mesh.

In order to estimate the error, we first make some estimates of the solutions u to the problem (3.8.1)–(3.8.3).

Lemma 3.8.1. *Let α be the interior angle at the point O, and $0 < \sigma < \frac{2\pi}{\alpha}$, then for every $t \in [0, T]$, $u = v + w$ on Ω, where v is the solution to*

$$\triangle v = 0, \tag{3.8.5}$$

$$v|_{\Gamma^*} = 0, \tag{3.8.6}$$

and $w \in H^2(\Omega)$. Moreover we have

$$\|u\|_{C([0,T];H^{2(\sigma)}(\Omega_0))} \leq C, \tag{3.8.7}$$

$$\|v\|_{L^\infty([0,T];H^{2(\sigma)}(\Omega))} \leq C, \tag{3.8.8}$$

$$\|w\|_{L^\infty([0,T];H^2(\Omega))} \leq C. \tag{3.8.9}$$

Proof. We study the operator $\triangle$ first. By [54] if $u_1 \in H^1(\Omega)$, u_1 satisfies the boundary condition (3.8.6), and $\triangle u_1 \in L^2(\Omega)$, then $u_1 = v_1 + w_1$, v_1 is a finite sum of functions with the form $r^{\frac{j\pi}{\alpha}} \sin \frac{j\pi}{\alpha}(\vartheta - \vartheta_0)$, therefore $v_1 \in H^{2(\sigma)}(\Omega)$, and w_1 belongs to $H^2(\Omega)$. Besides, [54] gives

$$\|v_1\|_{2(\sigma),\Omega} \leq C\|\triangle u_1\|_{0,\Omega},$$

$$\|w_1\|_{2,\Omega} \leq C\|\triangle u_1\|_{0,\Omega}.$$

Thus the domain of definition of the operator $\triangle$ can be given as

$$D(\triangle) = \{u \in H^{2(\sigma)}(\Omega_0) \bigcap H_0^1(\Omega_0); \triangle u \in L^2(\Omega_0)\},$$

and the mapping $\triangle : D(\triangle) \to L^2(\Omega_0)$ is surjective. Evidently $\overline{D(\triangle)} = L^2(\Omega_0)$. Let λ be a positive number, and v be the solution to

$$-\triangle v + \lambda v = g,$$

$$v\big|_{\partial\Omega_0} = 0.$$

We have the energy estimate

$$\lambda\|v\|_{0,\Omega_0} \leq \|g\|_{0,\Omega_0},$$

hence the resolvent of $\triangle$ satisfies

$$\|R(\lambda:\triangle)\| \leq \frac{1}{\lambda}.$$

By the Hille-Yosida Theorem we conclude that $\triangle$ is the infinitesimal generator of a C_0 semigroup of contractions, denote by $e^{t\triangle}, t \geq 0$. The solution to (3.8.1)–(3.8.3) can be expressed as

$$u(t) = e^{t\triangle}u_0 + \int_0^t e^{(t-\tau)\triangle} f \, d\tau.$$

Therefore we obtain [65]

$$\begin{aligned}\|u\|_{C^1([0,T];L^2(\Omega_0))} &\leq C,\\ \|u\|_{C([0,T];H^{2(\sigma)}(\Omega_0))} &\leq C.\end{aligned} \qquad (3.8.10)$$

To prove (3.8.8),(3.8.9) we rewrite the equation (3.8.1) in the form of

$$-\triangle u = f - \frac{\partial u}{\partial t},$$

and note that $\|f - \frac{\partial u}{\partial t}\|_{0,\Omega_0} \leq C$. We apply the result in [54] again, then (3.8.8) (3.8.9) follow. □

Let $\bar{u}_h(t)$ be an approximate function to the solution u for every $t \in [0,T]$ as follows: $\bar{u}_h(t)$ is the interpolation function of u by nodes on Ω^*. And $\bar{u}_h = w_h + v_h$ on Ω, where w_h is the interpolation function of w with respect to the finite element mesh, and $v_h \in V$, $v_h|_{\Gamma_0}$ is the interpolation function of v on Γ_0. Obviously $\bar{u}_h(t) \in S(\Omega_0)$.

Lemma 3.8.2. *Let $\bar{u}_h(t)$ be constructed as the above, then*

$$\|u(\cdot,t) - \bar{u}_h(t)\|_{0,\Omega_0} + h|u(\cdot,t) - \bar{u}_h(t)|_{1,\Omega_0} \leq Ch^2. \qquad (3.8.11)$$

Proof. By Theorem 3.3.1 and Theorem 3.3.2 we have

$$\|v - v_h\|_{0,\Omega} + h|v - v_h|_{1,\Omega} \leq Ch^2,$$

where we have used (3.8.8). Applying the estimate for interpolation operators [61],

$$\|u - \bar{u}_h\|_{0,\Omega^*} + h|u - \bar{u}_h|_{1,\Omega^*} \leq Ch^2,$$

$$\|w - w_h\|_{0,\Omega} + h|w - w_h|_{1,\Omega} \leq Ch^2,$$

where we have used (3.8.7),(3.8.9). Then (3.8.11) follows. □

Now we are in a position to estimate the error.

Theorem 3.8.1. If u and $u_h^{(m)}$ are the solutions to (3.8.1)–(3.8.3) and (3.8.4) respectively, then

$$\|u^{(m)} - u_h^{(m)}\|_{0,\Omega_0}^2 + \sum_{l=1}^{m} \|u^{(l)} - u_h^{(l)}\|_{1,\Omega_0}^2 \Delta t$$
$$\leq C\left(h^2 + \frac{h^4}{\Delta t^2} + \Delta t\right), \qquad m = 1, 2, \ldots,$$

where $u^{(m)} = u(\cdot, m\Delta t)$.

Proof. Let $\bar{u}_h^{(m)} = \bar{u}_h(m\Delta t), \eta^{(m)} = u^{(m)} - \bar{u}_h^{(m)}$, and $e^{(m)} = u^{(m)} - u_h^{(m)}$, then

$$\left(\frac{e^{(m+1)} - e^{(m)}}{\Delta t}, e^{(m+1)}\right) + a(e^{(m+1)}, e^{(m+1)})$$
$$= \left(\frac{e^{(m+1)} - e^{(m)}}{\Delta t}, \eta^{(m+1)}\right) + a(e^{(m+1)}, \eta^{(m+1)})$$
$$+ \left(\frac{u^{(m+1)} - u^{(m)}}{\Delta t}, \bar{u}_h^{(m+1)} - u_h^{(m+1)}\right) - \left(\frac{u_h^{(m+1)} - u_h^{(m)}}{\Delta t}, \bar{u}_h^{(m+1)} - u_h^{(m+1)}\right)$$
$$+ a(u^{(m+1)}, \bar{u}_h^{(m+1)} - u_h^{(m+1)}) - a(u_h^{(m+1)}, \bar{u}_h^{(m+1)} - u_h^{(m+1)}). \qquad (3.8.12)$$

The weak formulation of (3.8.1),(3.8.2) is

$$\left(\frac{\partial u^{(m)}}{\partial t}, v\right) + a(u^{(m)}, v) = (f, v), \qquad \forall v \in H_0^1(\Omega_0),$$

hence

$$a(u^{(m+1)}, \bar{u}_h^{(m+1)} - u_h^{(m+1)})$$
$$= (f, \bar{u}_h^{(m+1)} - u_h^{(m+1)}) - \left(\frac{\partial u^{(m+1)}}{\partial t}, \bar{u}_h^{(m+1)} - u_h^{(m+1)}\right).$$

Similarly (3.8.4) gives

$$a(u_h^{(m+1)}, \bar{u}_h^{(m+1)} - u_h^{(m+1)})$$
$$= (f, \bar{u}_h^{(m+1)} - u_h^{(m+1)}) - \left(\frac{u_h^{(m+1)} - u_h^{(m)}}{\Delta t}, \bar{u}_h^{(m+1)} - u_h^{(m+1)}\right).$$

Substituting them into (3.8.12), we have

$$\left(\frac{e^{(m+1)} - e^{(m)}}{\Delta t}, e^{(m+1)}\right) + a(e^{(m+1)}, e^{(m+1)}) = A_1^{(m+1)} + A_2^{(m+1)} + A_3^{(m+1)}, \qquad (3.8.13)$$

where
$$A_1^{(m+1)} = \left(\frac{e^{(m+1)} - e^{(m)}}{\Delta t}, \eta^{(m+1)}\right),$$

$$A_2^{(m+1)} = a(e^{(m+1)}, \eta^{(m+1)}),$$

$$A_3^{(m+1)} = \left(\frac{u^{(m+1)} - u^{(m)}}{\Delta t} - \frac{\partial u^{(m+1)}}{\partial t}, \bar{u}_h^{(m+1)} - u_h^{(m+1)}\right).$$

Upon summing (3.8.13) up with respect to m, we get

$$\sum_{l=0}^{m}(e^{(l+1)} - e^{(l)}, e^{(l+1)}) + \sum_{l=0}^{m} a(e^{(l+1)}, e^{(l+1)})\Delta t$$
$$= \sum_{l=0}^{m}(A_1^{(l+1)} + A_2^{(l+1)} + A_3^{(l+1)})\Delta t. \qquad (3.8.14)$$

Let us estimate the terms on the right in turn. For the first term we have

$$\sum_{l=0}^{m} A_1^{(l+1)}\Delta t = -\sum_{l=1}^{m}(e^{(l)}, \eta^{(l+1)} - \eta^{(l)}) + (e^{(m+1)}, \eta^{(m+1)}) - (e^{(0)}, \eta^{(1)})$$
$$\leq \sum_{l=1}^{m} \|e^{(l)}\|_0 \|\eta^{(l+1)} - \eta^{(l)}\|_0 + \|e^{(m+1)}\|_0 \|\eta^{(m+1)}\|_0 + \|e^{(0)}\|_0 \|\eta^{(1)}\|_0$$
$$\leq Ch^2 \left(\sum_{l=1}^{m} \|e^{(l)}\|_0 + \|e^{(m+1)}\|_0 + \|e^{(0)}\|_0\right),$$

where we have applied (3.8.11). From the Poincare inequality we have

$$\|e^{(l)}\|_0^2 \leq Ca(e^{(l)}, e^{(l)}),$$

hence for $\varepsilon > 0$,

$$h^2 \sum_{l=1}^{m+1} \|e^{(l)}\|_0 \leq \frac{1}{2}\sum_{l=1}^{m+1}\left(\frac{h^4}{\varepsilon \Delta t} + \varepsilon \|e^{(l)}\|_0^2 \Delta t\right)$$
$$\leq \frac{m+1}{2}\frac{h^4}{\varepsilon \Delta t} + C\varepsilon \sum_{l=1}^{m+1} a(e^{(l)}, e^{(l)})\Delta t.$$

Set $\varepsilon C = \frac{1}{8}$, then

$$\sum_{l=0}^{m} A_1^{(l+1)}\Delta t \leq \frac{1}{8}\sum_{l=1}^{m+1} a(e^{(l)}, e^{(l)})\Delta t + \frac{Ch^4}{\Delta t^2} + Ch^2\|e^{(0)}\|_0, \qquad (3.8.15)$$

where we have noticed $m \leq \frac{T}{\Delta t}$. For the second term of (3.8.14) we have

$$\sum_{l=0}^{m} A_2^{(l+1)} \Delta t \leq \sum_{l=0}^{m} |e^{(l+1)}|_1 |\eta^{(l+1)}|_1 \Delta t$$

$$\leq \frac{1}{8} \sum_{l=0}^{m} |e^{(l+1)}|_1^2 \Delta t + C \sum_{l=0}^{m} |\eta^{(l+1)}|_1^2 \Delta t$$

$$\leq \frac{1}{8} \sum_{l=0}^{m} a(e^{(l+1)}, e^{(l+1)}) \Delta t + Ch^2, \qquad (3.8.16)$$

where we have applied (3.8.11). For the third term of (3.8.14) we have

$$\sum_{l=0}^{m} A_3^{(l+1)} \Delta t = \sum_{l=0}^{m} \int_{l\Delta t}^{(l+1)\Delta t} \left(\frac{\partial u}{\partial t} - \frac{\partial u^{(l+1)}}{\partial t}, \bar{u}_h^{(l+1)} - u_h^{(l+1)} \right) dt$$

$$= \sum_{l=0}^{m} \int_{l\Delta t}^{(l+1)\Delta t} \left(\frac{\partial u}{\partial t} - \frac{\partial u^{(l+1)}}{\partial t}, \bar{u}_h^{(l+1)} - u^{(l+1)} \right) dt$$

$$+ \sum_{l=0}^{m} \int_{l\Delta t}^{(l+1)\Delta t} \left(\frac{\partial u}{\partial t} - \frac{\partial u^{(l+1)}}{\partial t}, u^{(l+1)} - u_h^{(l+1)} \right) dt$$

$$\equiv B_1 + B_2.$$

Applying (3.8.10),(3.8.11) we get

$$B_1 \leq \sum_{l=0}^{m} \int_{l\Delta t}^{(l+1)\Delta t} C \|\bar{u}_h^{(l+1)} - u^{(l+1)}\|_0 \, dt$$

$$\leq Ch^2.$$

Applying (3.8.1) we get

$$B_2 = \sum_{l=0}^{m} \int_{l\Delta t}^{(l+1)\Delta t} (\triangle u - \triangle u^{(l+1)}, e^{(l+1)}) \, dt$$

$$= \sum_{l=0}^{m} \int_{l\Delta t}^{(l+1)\Delta t} a(u^{(l+1)} - u, e^{(l+1)}) \, dt$$

$$\leq \sum_{l=0}^{m} \int_{l\Delta t}^{(l+1)\Delta t} \left\{ \frac{1}{8} a(e^{(l+1)}, e^{(l+1)}) + Ca(u^{(l+1)} - u, u^{(l+1)} - u) \right\} dt.$$

The equation (3.8.1) and the inequality (3.8.10) imply that

$$a(u^{(l+1)} - u, u^{(l+1)} - u) = -(\triangle u^{(l+1)} - \triangle u, u^{(l+1)} - u)$$

$$= \left(\frac{\partial u^{(l+1)}}{\partial t} - \frac{\partial u}{\partial t}, u^{(l+1)} - u \right) \leq C \|u^{(l+1)} - u\|_0.$$

Therefore

$$B_2 \le \frac{1}{8}\sum_{l=0}^{m} a(e^{(l+1)}, e^{(l+1)})\Delta t + C\sum_{l=0}^{m} \int_{l\Delta t}^{(l+1)\Delta t} \|u^{(l+1)} - u\|_0 \, dt.$$

We apply (3.8.10) again and obtain

$$\|u^{(l+1)} - u\|_0 = \left\|\int_t^{(l+1)\Delta t} \frac{\partial u(\cdot,\tau)}{\partial \tau} d\tau\right\|_0$$
$$\le \int_t^{(l+1)\Delta t} \left\|\frac{\partial u(\cdot,\tau)}{\partial \tau}\right\|_0 d\tau \le C\Delta t,$$

therefore

$$B_2 \le \frac{1}{8}\sum_{l=0}^{m} a(e^{(l+1)}, e^{(l+1)})\Delta t + C\Delta t.$$

Consequently

$$\sum_{l=0}^{m} A_3^{(l+1)} \le \frac{1}{8}\sum_{l=0}^{m} a(e^{(l+1)}, e^{(l+1)})\Delta t + Ch^2 + C\Delta t. \tag{3.8.17}$$

We substitute (3.8.15),(3.8.16),(3.8.17) into (3.8.14) and obtain

$$\sum_{l=0}^{m}(e^{(l+1)} - e^{(l)}, e^{(l+1)}) + \frac{1}{2}\sum_{l=0}^{m} a(e^{(l+1)}, e^{(l+1)})\Delta t$$
$$\le Ch^2 + C\frac{h^4}{\Delta t^2} + C\Delta t + Ch^2\|e^{(0)}\|_0. \tag{3.8.18}$$

Since

$$|(e^{(l)}, e^{(l+1)})| \le \frac{1}{2}(\|e^{(l)}\|_0^2 + \|e^{(l+1)}\|_0^2),$$

we have

$$\sum_{l=0}^{m}(e^{(l+1)} - e^{(l)}, e^{(l+1)}) \le \frac{1}{2}\|e^{(m+1)}\|_0^2 - \frac{1}{2}\|e^{(0)}\|_0^2.$$

Recall that we take $u_h^{(0)} = \Pi u_0$, so by the estimates of interpolation functions we get

$$\|e^{(0)}\|_0 = \|u^{(0)} - u_h^{(0)}\|_0 = \|u_0 - \Pi u_0\|_0 \le Ch^2|u_0|_2.$$

We obtain from (3.8.18) that

$$\|e^{(m+1)}\|_0^2 + \frac{1}{2}\sum_{l=0}^{m} a(e^{(l+1)}, e^{(l+1)})\Delta t \le C\left(h^2 + \frac{h^4}{\Delta t^2} + \Delta t\right).$$

Noting that
$$\|e^{(l+1)}\|_1^2 \leq Ca(e^{(l+1)}, e^{(l+1)}),$$
we finally reach the conclusion. $\square$

3.9 Variational inequalities (II)

In this section the domain and mesh is also the same as those in §3.6, and the problem is
$$u \in U, \quad \text{and} \quad J(u) = \min_{v \in U} J(v), \tag{3.9.1}$$
where
$$J(u) = \frac{1}{2}a(u,u) - (f,u), \tag{3.9.2}$$
and $f \in L^2(\Omega_0)$, and
$$U = \{u \in H_0^1(\Omega_0); u \geq \varphi \quad \text{a.e. in } \Omega_0\}. \tag{3.9.3}$$

We assume that the "obstacle" function φ satisfies: $\varphi \in H^2(\Omega_0), \varphi|_{\partial\Omega_0} \leq 0$, and $\triangle\varphi \geq 0$.

The problem (3.9.1)–(3.9.3) is equivalent to [61]: find $u \in U$, such that
$$a(u, v-u) \geq (f, v-u), \quad \forall v \in U. \tag{3.9.4}$$

Let $S(\Omega_0)$ be the infinite element subspace of $H_0^1(\Omega_0)$ given in §3.8. We construct a subset U_h of it, which is the approximation to U.
$$U_h = \left\{ u \in S(\Omega_0); \int_{e_i} u(x)\,dx \geq \int_{e_i} \varphi(x)\,dx \text{ for all triangular elements } e_i \right.$$
$$\left. \text{associated with } U^* \text{ or } W \right\}.$$

The infinite element solution u_h satisfies: $u_h \in U_h$, and
$$a(u_h, v - u_h) \geq (f, v - u_h) \quad \forall v \in U_h. \tag{3.9.5}$$

We assume that the meshes satisfy, besides the conditions (A),(B), the condition

(J) Each element associated with U^* or W possesses at least one node in the domain Ω_0.

In order to estimate the error, we first make some estimates of the solutions u to the problem (3.9.1)–(3.9.3).

3.9 VARIATIONAL INEQUALITIES (II)

Lemma 3.9.1. There is a unique $u_\varepsilon \in U$ for any $\varepsilon > 0$ such that $u_\varepsilon \in U$, $\triangle u_\varepsilon \in L^2(\Omega_0)$, and
$$u_\varepsilon - \varepsilon \triangle u_\varepsilon = u, \tag{3.9.6}$$
where u is the solution to the problem (3.9.1)–(3.9.3).

Proof. By the Lax-Milgram Theorem, we can prove there is a unique solution to the problem: find $u_\varepsilon \in H_0^1(\Omega_0)$, such that
$$\varepsilon a(u_\varepsilon, v) + (u_\varepsilon, v) = (u, v), \qquad \forall v \in H_0^1(\Omega_0).$$
Therefore
$$\triangle u_\varepsilon = \frac{1}{\varepsilon}(u_\varepsilon - u) \in L^2(\Omega_0).$$
Let us prove that $u_\varepsilon \in U$, thus u_ε is the desired solution. We define
$$(\varphi - u_\varepsilon)^+ = \max\{\varphi - u_\varepsilon, 0\},$$
then $(\varphi - u_\varepsilon)^+ \in H_0^1(\Omega_0)$ [52]. We have
$$-(\triangle(\varphi - u_\varepsilon), (\varphi - u_\varepsilon)^+) = a((\varphi - u_\varepsilon)^+, (\varphi - u_\varepsilon)^+) \geq 0.$$
Therefore by (3.9.6)
$$(\triangle \varphi, (\varphi - u_\varepsilon)^+) = (\triangle u_\varepsilon + \triangle(\varphi - u_\varepsilon), (\varphi - u_\varepsilon)^+)$$
$$\leq (\triangle u_\varepsilon, (\varphi - u_\varepsilon)^+)$$
$$= \left(\frac{1}{\varepsilon}(u_\varepsilon - u), (\varphi - u_\varepsilon)^+\right).$$
On the other hand by the assumption
$$(\triangle \varphi, (\varphi - u_\varepsilon)^+) \geq 0,$$
hence
$$(u_\varepsilon - u, (\varphi - u_\varepsilon)^+) \geq 0.$$
We define a sebset of Ω_0, $E = \{x \in \Omega_0; \varphi(x) \geq u_\varepsilon(x)\}$, then
$$\int_E (u_\varepsilon - u)(\varphi - u_\varepsilon)\, dx \geq 0.$$
Since $u(x) \geq \varphi(x)$ holds a.e. on Ω_0, $u(x) - u_\varepsilon(x) \geq 0$ a.e. on E, thus
$$\int_E (u_\varepsilon - u)(\varphi - u_\varepsilon)\, dx \leq 0.$$
Consequently
$$\int_E (u_\varepsilon - u)(\varphi - u_\varepsilon)\, dx = 0,$$
which implies
$$(u_\varepsilon - u)(\varphi - u_\varepsilon) = 0, \quad \text{a.e. } x \in E. \tag{3.9.7}$$
On the subset of E where $u_\varepsilon(x) = u(x)$ we have $u_\varepsilon(x) = \varphi(x) = u(x)$ a.e., and on the subset of E where $u_\varepsilon(x) < u(x)$, we have by (3.9.7) that $\varphi(x) = u_\varepsilon(x)$ a.e., hence
$$\varphi(x) = u_\varepsilon(x), \quad \text{a.e. } x \in E.$$
According to the definition, $u_\varepsilon(x) > \varphi(x)$ on $\Omega_0 \setminus E$. Thus $u_\varepsilon \in U$ and the proof is complete. □

Lemma 3.9.2. *The solution to (3.9.4) can be decomposed on Ω as $u = v + w$, where v is the solution to (3.8.5),(3.8.6) and $w \in H^2(\Omega)$.*

Proof. From the equation (3.9.4) and Lemma 3.9.1 we have
$$a(u, u_\varepsilon - u) \geq (f, u_\varepsilon - u).$$
Since
$$a(u_\varepsilon - u, u_\varepsilon - u) \geq 0,$$
we get
$$a(u_\varepsilon, u_\varepsilon - u) \geq a(u, u_\varepsilon - u),$$
therefore
$$a(u_\varepsilon, u_\varepsilon - u) \geq (f, u_\varepsilon - u).$$
From (3.9.6) we get
$$a(u_\varepsilon, \varepsilon \triangle u_\varepsilon) \geq (f, \varepsilon \triangle u_\varepsilon).$$
Integrating by parts we obtain
$$a(u_\varepsilon, \varepsilon \triangle u_\varepsilon) = -\varepsilon(\triangle u_\varepsilon, \triangle u_\varepsilon),$$
which leads to
$$\|\triangle u_\varepsilon\|_0^2 \leq \|f\|_0 \|\triangle u_\varepsilon\|_0,$$
or
$$\|\triangle u_\varepsilon\|_0 \leq \|f\|_0. \tag{3.9.8}$$
Applying (3.9.6) and (3.9.8) we conclude that $u_\varepsilon \to u$ as $\varepsilon \to 0$ in $L^2(\Omega_0)$. Therefore
$$a(u_\varepsilon, u_\varepsilon) = -(\triangle u_\varepsilon, u_\varepsilon) \leq C,$$
which yields
$$\|u_\varepsilon\|_1 \leq C.$$
Consequently we can extract a sequence $\{u_{\varepsilon_l}\}_{l=1}^\infty$, such that $\varepsilon_l \to 0$ $(l \to \infty)$ and u_{ε_l} converges to u weakly as $l \to \infty$ in $H^1(\Omega_0)$. (3.9.8) also implies that we can extract a subsequence, still denoted by $\{u_{\varepsilon_l}\}$, such that $\triangle u_{\varepsilon_l}$ converges weakly. Let $g \in L^2(\Omega_0)$ be the limit. For $v \in H_0^1(\Omega_0)$ we have
$$a(u_{\varepsilon_l}, v) = -(\triangle u_{\varepsilon_l}, v) \to -(g, v),$$
and
$$a(u_{\varepsilon_l}, v) \to a(u, v),$$
therefore
$$a(u, v) = (g, v). \tag{3.9.9}$$
The results in [54] lead to the conclusion. $\square$

Following §3.8, we construct an approximate function $\bar{u}_h$ to the solution u, then Lemma 3.8.2 can be applied here. We have
$$\|u - \bar{u}_h\|_{0,\Omega_0} + h|u - \bar{u}_h|_{1,\Omega_0} \leq Ch^2. \tag{3.9.10}$$
Inequality (3.9.10) is essential for the error estimates.

3.9 VARIATIONAL INEQUALITIES (II)

Lemma 3.9.3. *The set U_h is non-empty.*

Proof. Let χ be a continuous function on Ω_0, such that $\chi \in U^*$ on Ω^* and $\chi \in W$ on Ω, and $\chi(x) \equiv \alpha$ on all interior nodes, where α is a constant to be determined.

Let e_i be an element associated with U^* or W. We notice the condition (J), then we have by (3.9.10)

$$\int_{e_i} (\bar{u}_h + \chi)\, dx = \int_{e_i} (u + \chi)\, dx + \int_{e_i} (\bar{u}_h - u)\, dx$$

$$\geq \int_{e_i} (u + \chi)\, dx - \left(\int_{e_i} dx\right)^{\frac{1}{2}} \left(\int_{e_i} (\bar{u}_h - u)^2\, dx\right)^{\frac{1}{2}}$$

$$\geq \int_{e_i} u\, dx + \frac{\alpha}{3}\, \text{meas}\, e_i - Ch^3$$

$$\geq \int_{e_i} u\, dx + \left(\frac{\alpha}{3} - Ch\right)\, \text{meas}\, e_i.$$

We take $\alpha = 3Ch$, than apply (3.9.3) to get

$$\int_{e_i} (\bar{u}_h + \chi)\, dx \geq \int_{e_i} \varphi\, dx.$$

Therefore $\bar{u}_h + \chi \in U_h$. $\square$

The following error estimate is given in [61] (Theorem 5.1.1):

$$\|u - u_h\|_1^2 \leq C(\inf_{\eta \in U_h} \{\|u - \eta\|_1^2 + \|\triangle u + \varphi\|_0 \|u - \eta\|_0\}$$
$$+ \|\triangle u + f\|_0 \inf_{\psi \in U} \|u_h - \psi\|_0). \tag{3.9.11}$$

To estimate the right hand side of (3.9.11), we need the following lemma:

Lemma 3.9.4. *If e_i is one triangular element, and a linear function l satisfies $|l| \leq r$ on e_i, then*

$$|l|_{1,e_i}^2 \leq Cr^2.$$

The proof of it is straightforward, thus omitted here.

Theorem 3.9.1. *If u and u_h are solutions to (3.9.4),(3.9.5) respectively, then*

$$\|u - u_h\|_{1,\Omega_0} \leq Ch^{\frac{1}{2}}. \tag{3.9.12}$$

Proof. By (3.9.9), $\triangle u = -g$, hence

$$\|\triangle u + f\|_0 \leq C.$$

Let χ be the function introduced in Lemma 3.9.3, then we have

$$\inf_{\eta \in U_h} \{\|u - \eta\|_1^2 + \|\triangle u + f\|_0 \|u - \eta\|_0\}$$
$$\leq \|u - \bar{u}_h - \chi\|_1^2 + C\|u - \bar{u}_h - \chi\|_0. \qquad (3.9.13)$$

The number of elements neighboring $\partial\Omega_0$ associated with U^* and W is $O(\frac{1}{h})$. Using Lemma 3.9.4 we get

$$\|\chi\|_{1,\Omega_0}^2 \leq C(3Ch)^2 \frac{1}{h} = Ch.$$

Therefore by (3.9.10),

$$\|u - \bar{u}_h - \chi\|_1^2 \leq Ch^2 + Ch \leq Ch,$$
$$\|u - \bar{u}_h - \chi\|_0 \leq Ch^2 + Ch \leq Ch.$$

Substituting them into (3.9.13) we get

$$\inf_{\eta \in U_h} \{\|u - \eta\|_1^2 + \|\triangle u + f\|_0 \|u - \eta\|_0\} \leq Ch. \qquad (3.9.14)$$

We turn now to estimate the last term in (3.9.11). We define an interpolation operator Π in $H^1(\Omega_0)$ as follows:

$$\Pi v = \frac{1}{\operatorname{meas} e_i} \int_{e_i} v(t)\,dt, \quad \forall x \in e_i, \forall v \in H^1(\Omega_0),$$

where as before e_i is an element associated with U^* or W. Besides we define a set $E = \{x \in \Omega_0; u_h(x) < \varphi(x)\}$. And we have on E that

$$0 < \varphi - u_h = \varphi - \Pi\varphi + \Pi\varphi - \Pi u_h + \Pi u_h - u_h$$
$$\leq \varphi - \Pi\varphi + \Pi u_h - u_h,$$

where we have noticed $\Pi u_h \geq \Pi \varphi$. Let $v_1 = \max\{u_h, \varphi\}$, then $v_1 \in U$, hence

$$\inf_{\psi \in U} \|u_h - \psi\|_0 \leq \|u_h - v_1\|_0 = \left(\int_E (\varphi - u_h)^2\,dx\right)^{\frac{1}{2}}$$
$$\leq \left(\int_E (\varphi - \Pi\varphi)^2\,dx\right)^{\frac{1}{2}} + \left(\int_E (u_h - \Pi u_h)^2\,dx\right)^{\frac{1}{2}}$$
$$\leq \|\varphi - \Pi\varphi\|_{0,\Omega_0} + \|u_h - \Pi u_h\|_{0,\Omega_0}.$$

Applying the approach in [61] to estimate the interpolation error, we can get

$$\|\varphi - \Pi\varphi\|_{0,\Omega_0} \leq Ch\|\varphi\|_{1,\Omega_0},$$
$$\|u_h - \Pi u_h\|_{0,\Omega_0} \leq Ch\|u_h\|_{1,\Omega_0}.$$

Thus
$$\inf_{\psi \in U} \|u_h - \psi\|_0 \leq Ch + Ch\|u_h\|_{1,\Omega_0}. \qquad (3.9.15)$$

We substitute (3.9.14),(3.9.15) into (3.9.11) and get

$$\|u - u_h\|_1^2 \leq Ch + Ch\|u_h\|_1$$
$$\leq Ch + Ch\|u - u_h\|_1 + Ch\|u\|_1.$$

From (3.9.9) we obtain
$$\|u\|_1 \leq C\|g\|_0,$$
hence
$$\|u - u_h\|_1^2 \leq Ch + \frac{1}{2}(\|u - u_h\|_1^2 + C^2 h^2) + Ch\|g\|_0,$$
consequently
$$\|u - u_h\|_1^2 \leq Ch,$$
which gives (3.9.12) and completes the proof. □

Notes to Chapter 3

The material of §3.1 is based on [23] and [25], and the material of §3.2, §3.3 is based on [6],[23] and [25]. The material of §3.4 and §3.5 is mainly based on [23],[25], but some lemmas there are the propositions in [48],[57] and [64]. The material of §3.6 is mainly based on [6],[19] with some significant changes. There were some gaps in the original proof, now it becomes rigorous. The material of §3.7 is based on [34] and [67], and the material of §3.8 and §3.9 is based on [39] and [41] respectively.

In the references some theorems are stated and proved in accordance with plane elasticity problems. For reading convenience we change them to the Laplace equation. The results in §3.3, §3.6–§3.9 are valid for many elliptic equations. The conclusions and argument in §3.4, §3.5 seems much limited. We still do not know to what extent they could be generalized.

4 Examples

4.1 A glance at the error

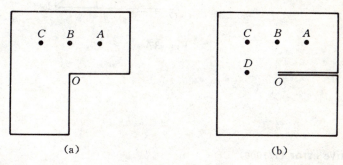

Fig. 36

Two simple domains are shown in Fig.36. The domain (a) is a so called L-shape domain with a reentrant interior angle $\alpha = \frac{3\pi}{2}$. The domain (b) has a crack with a reentrant interior angle $\alpha = 2\pi$. It is easy to obtain two particular solutions to the Laplace equation on these domains,

(a) $$u = r^{\frac{2}{3}} \sin\left(\frac{2}{3}\vartheta\right),\qquad(4.1.1)$$

(b) $$u = r^{\frac{1}{2}} \sin\left(\frac{1}{2}\vartheta\right),\qquad(4.1.2)$$

where (r, ϑ) are the polar coordinates. They vanish along the adjacent sides to the point O on the two domains respectively.

We take the values of u on the boundary as boundary data and solve these boundary value problems by means of the infinite element method. The mesh for the domain (a) is shown in Fig.37. By symmetry of the domain and of the boundary data, only half of the domain is taken into account. Here we apply a quite coarse mesh, since our aim is to show the infinite element method can give satisfactory results even for this extreme case. We take $\xi = 0.5, 0.6, 0.7, 0.8, 0.9$, and compare the approximate solutions with exact solutions u at points A, B, C, D. The largest relative errors are shown in Fig.38.

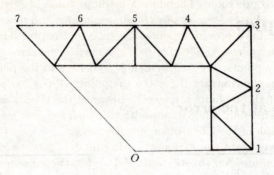

Fig. 37

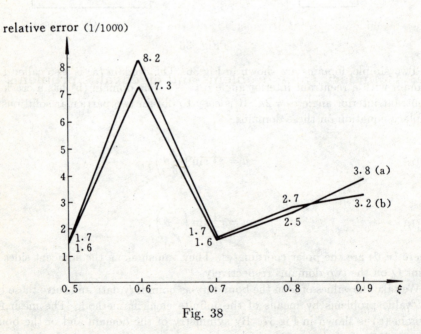

Fig. 38

Let us make an analysis of the curves shown in Fig.38. Too small and too large values of ξ are not satisfactory. For the former case the smaller the parameter ξ, the coarser the mesh, then the interpolation error is larger. For the later case although the mesh is refined in one direction, but it has no change in the other direction. The smallest interior angle of the triangular elements decreases as ξ increases, which is disadvantageous to precision. Therefore a medium value of ξ is appropriate. Now we should explain why the curves possess abnormal ups and downs between $\xi = 0.5$ and $\xi = 0.7$. We have proved superconvergence in Chapter Three. If $\xi = 0.5$ it happened that the points A, B, C, D are nodes, hence the error at these points are smaller than other points. If $\xi = 0.7$, then $\xi^2 = 0.49$, and the points A, B, C, D are near to nodes. On the contrary if $\xi = 0.6$ the points A, B, C, D are far from nodes, hence the error at those points are larger. However the results are very good since the relative errors are only a few thousandths for such a coarse mesh. Besides Fig.38 offers a visual graph of the effect of superconvergence.

Next let us investigate the exponents. Taking $\xi = 0.6$, we compare the exact values with the infinite element solutions and obtain the following:

Table 3

(a)	exact solution	2/3	2	10/3	14/3	6
	infinite element	0.667	1.954	3.205	4.242	6.668
(b)	exact solution	1/2	3/2	5/2	7/2	9/2
	infinite element	0.500	1.495	2.463	3.339	4.150

They are in good agreement, especially for the low order terms.

4.2 Boundary value problems and eigenvalue problems

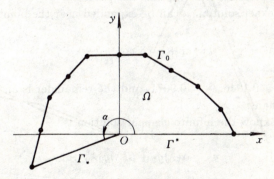

Fig. 39

Fig.39 shows a polygonal domain Ω with a reentrant interior angle $\alpha > \pi$ at the point O. We aim at finding the numerical solution to the following boundary value problem:

$$\triangle u + \lambda u = 0, \qquad (4.2.1)$$

$$\left.\frac{\partial u}{\partial \nu}\right|_{\Gamma^*} = \left.\frac{\partial u}{\partial \nu}\right|_{\Gamma_*} = 0, \qquad (4.2.2)$$

$$u|_{\Gamma_0} = f. \tag{4.2.3}$$

Let $\lambda \geq 0$. There are 11 nodes with the following coordinates (Table 4) taken along the boundary Γ_0:

Table 4

	1	2	3	4	5	6	7	8	9	10	11
x	4.0	3.5	2.7	1.85	1.0	0	-1.2	-1.8	-2.4	-2.7	-3.0
y	0	0.8	1.6	2.1	2.6	2.6	2.6	1.95	1.3	0.15	-1.0

An infinite and similar triangulation is established over the domain as Fig.10. It is easy to obtain the expansion of the solution near the point O,

$$u = a_0 J_0(\sqrt{\lambda} r) + \sum_{j=1}^{\infty} a_j J_{\frac{j\pi}{\alpha}}(\sqrt{\lambda} r) \cos \frac{j\pi}{\alpha} \vartheta \tag{4.2.4}$$

by means of separation of variables. The term associated with $j = 1$ is selected to be a model solution,

$$u = J_{\frac{\pi}{\alpha}}(\sqrt{\lambda} r) \cos \frac{\pi}{\alpha} \vartheta. \tag{4.2.5}$$

We set $\lambda = 0.1$. Nodal values on Γ_0 are determined by (4.2.5), then the above problem is solved by the infinite element method. Comparing it with the exact solution we analyze the numerical result as follows:

(a) Asymptotic behavior near the singular point.

By (4.2.5) the exact solution u can be expanded near the point O as

$$u = c_1 + c_2 r^{\frac{\pi}{\alpha}} \cos \frac{\pi}{\alpha} \vartheta + \cdots, \tag{4.2.6}$$

where $c_1 = 0$, $c_2 = 0.1946$, $\frac{\pi}{\alpha} = 0.9071$, and the remainder is an infinitesimal with an order higher than $r^{\frac{\pi}{\alpha}}$.

By (1.14.6) we know the infinite element solution is

$$y_k = \alpha(\lambda) g_1 + \beta(\lambda) g_2 \lambda_2^k + \cdots \tag{4.2.7}$$

near the point O. Applying an argument on the eigenvalues of the transfer matrix X as §1.9, we get

$$\frac{\log \lambda_2}{\log \xi} \approx \frac{\pi}{\alpha}.$$

Using the approach of §1.9, the approximate value to c_2^2, denoted by $\tilde{c}_2^2$, is obtained upon comparing the strain energy of the second terms of (4.2.6) and (4.2.7). The results are illustrated in the following table:

Table 5

ξ	$\frac{\log \lambda_2}{\log \xi}$	relative error	$\tilde{c}_2^2$	relative error
0.80	0.9083	0.13%	0.03728	1.59%
0.85	0.9084	0.14%	0.03731	1.52%
0.90	0.9087	0.18%	0.03714	1.97%
0.95	0.9094	0.25%	0.03717	1.86%

(b) Comparison of nodal values.

We evaluate nodal values at the seventh, tenth, twelfth and twenty fifth layers for $\xi = 0.80, 0.85, 0.90, 0.95$ respectively, and compare them with exact values. The relative errors are $1.09\%, 1.21\%, 1.59\%, 2.8\%$.

We turn now to set $f = 0$ in (4.2.3) and consider the eigenvalue problem (4.2.1), (4.2.2),(4.2.3). For the sake of easily getting exact solution, the domain is selected to be a sector,
$$\Omega = \{(r, \vartheta); 0 < r < R, 0 < \vartheta < \alpha\}.$$
11 nodes are taken along the circle $r = R$ too. Denote by Γ_0 the linking line of the nodes, and by Ω_h the domain surrounded by Γ_0, Γ^* and Γ_*, cf Fig.40.

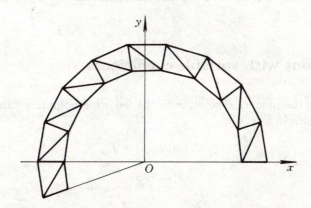

Fig. 40

By (4.2.4) it is clear that the smallest eigenvalue λ is the smallest zero point of
$$J_0(\sqrt{\lambda}R) = 0.$$
If $R = 4$, then
$$\lambda = 0.3614.$$
The approximate eigenvalues μ of the infinite element solutions are:

Table 6

ξ	0.65	0.75	0.85
μ	0.3748	0.3769	0.3820

Compared with the exact value, the approximate values are systematically bigger. There are two factors causing this inclination. The first one is discretization, and the second one is $\Omega_h \subset \Omega$.

Let us examine the influence of the second factor. We define a similarity transformation for the entire combined element Ω_h with center O and the constant of proportionality $1/\cos 10°$, then the image Γ'_0 of Γ_0 is exactly the external tangent polygon of the circle $r = R$. Denote by μ' the smallest eigenvalue with respect to this domain, then by the argument of Theorem 2.12.5,

$$\mu' = \mu \cos^2 10°.$$

According to this formula we get

Table 7

ξ	0.65	0.75	0.85
μ'	0.3634	0.3655	0.3705

We discover that precision is improved. It is because in evaluating μ' discretization makes the approximate eigenvalue increase, but the expansion of domain makes it decrease. The effect of cancellation makes the approximate value come close to the exact one.

4.3 Equations with variable coefficients

We calculate the numerical solution to the following boundary value problem on the domain shown in Fig.36(a),

$$-\nabla(a(x)\nabla u) = f, \tag{4.3.1}$$

$$u\big|_{\partial\Omega} = g. \tag{4.3.2}$$

We take

$$a(x) = 1 + \frac{r^2}{8},$$

then

$$-\nabla(u(0)\nabla) = -\triangle.$$

Besides, we take

$$u = r^{\frac{2}{3}} \sin\left(\frac{2}{3}\vartheta\right) + \frac{1}{8}x_1^2 x_2^2,$$

where the first term is from (4.1.1). The second term is added deliberately, such that u is not a solution to the Laplace equation. Then

$$f = -\frac{1}{6}r^{\frac{2}{3}}\cos\left(\frac{4\vartheta}{3}\right) - \frac{r^2}{4}\left(1 + \frac{r^2}{8}\right) - \frac{x_1^2 x_2^2}{8}.$$

The infinite element mesh is shown in Fig.37, and the finite element mesh is shown in Fig.41.

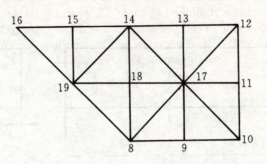

Fig. 41

We take
$$\xi = 0.7.$$

The solutions are (in the order of nodes):

$$0.0000, 0.1796, 0.3838, 0.3822, 0.4440,$$
$$0.5505, 0.7052, 0.0000, 0.0000, 0.0000,$$
$$0.1793, 0.3712, 0.3739, 0.4200, 0.5333,$$
$$0.6797, 0.1900, 0.2562, 0.4018.$$

Comparing the sum of two mesh functions with the exact solution at the points $17, 18, 19$, we obtain the errors (exact solution minus approximate one):

$$0.0018, 0.0046, -0.0173.$$

The range of values of the exact solution is

$$[0, 2^{\frac{1}{3}} + \tfrac{1}{8}] \approx [0, 1.3849].$$

Regarding the maximum error and the maximum value, the relative error is about 1.2%.

In carrying out the numerical integration, the traperoid formula with recurrent refinement is applied. The controlling error is 10^{-4} and the controlling time of recurrence is 7. However the error for some integration can not still be as small as 10^{-4}. The maximum error is 1.89×10^{-3}. If we improve the accuracy of the numerical integration, the precision of the solution might be even higher.

4.4 Stress intensity factors

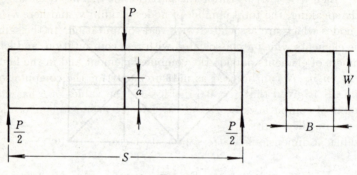

Fig. 42

Fig.42 shows a three points bending specimen for fracture experiments. There is a crack a in length sited at the middle of the specimen. As the force of action P increases gradually, the deformation of the specimen increases until the crack expands and the specimen collapses finally. We evaluate the stress intensity factors K_I for different P, then in coordination with the experiment we can get the critical value of K_I, i.e. K_{Ic}, which serves as a standard of mechanical design.

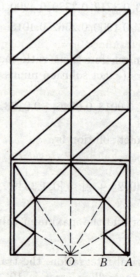

Fig. 43

K_I is evaluated by the infinite element method. The above problem satisfies the symmetric condition of Fig.11, hence to solve half of the domain is sufficient. Fig.43

shows a typical partition, where linear triangular elements are applied, point O is the crack tip and OA is the crack. Surrounded in the double line is the combined element Ω. if the number of nodes on Γ_0 is denoted by n, and the total number of nodes by n_1, then $n = 7, n_1 = 16$ for the partition of Fig.43. It should be noticed that, strictly speaking, the total number of nodes is infinity, and here n_1 means the number of nodes which are associated with real computation. In the same way, we consider the elements which are associated with real computation, and let N be the sum of numbers of elements outside the combined element and in the first layer Ω_1, then $N = 24$. n, n_1, N can be used as measurements for the computation work of this problem. It is given that $\xi = 0.6$ in Fig.43. The values of ξ has no influence on the amount of computation work, but we have seen and we will see later it has great influence on the results.

We introduce a dimensionless stress intensity factor,

$$K_I^* = \frac{K_I B W^{\frac{3}{2}}}{M},$$

where $M = \frac{1}{4} Ps$ stands for the bending moment. Let the Poisson ratio be 0.3 and $\frac{s}{W} = 4$, $\frac{a}{W} = 0.5$, then with the partition of Fig.43, different K_I^* for different ξ are as follows:

Table 8

ξ	0.6	0.7	0.8	0.85	0.9
K_I^*	9.876	10.233	10.366	10.397	10.350

It is known that the strain energy of the finite element solution is smaller than the exact value, nevertheless the solutions of the infinite element method possess the same property. Now K_I is determined from the strain energy, therefore we realize that the K_I^* value for $\xi = 0.85$ is closer to the exact one.

From the above results we see that precision increases as ξ, but if ξ exceeds a certain extent, the increasing of ξ has no significant influence to the result, or even makes precision decrease. We have seen in §4.1 that to obtain satisfactory results, the values of n and ξ should be adjust to each other.

The value of K_I is independent of the Poisson ratio for plane elasticity problems [51], but the strain energy in the infinite element method is approximate, therefore the value of the Poisson ratio has influence on K_I. For the sake of contrast we take two values of the Poisson ratio, 0.2 and 0.3, and numerical results show that the relative error is only 0.5%. Hence the Poisson ratio has no significant influence.

Denoting by $\lambda_1, \lambda_2, \ldots$ the eigenvalues of the transfer matrix X, we evaluate the exponents in accordance with the formula

$$\alpha_j = \frac{\log \lambda_j}{\log \xi}.$$

Let us observe the changing of exponents with respect to a series of refining meshes, it is obtained that:

Table 9

n	ξ	α_0	α_1	α_2	α_3	α_4
7	0.85	0	0.5199	1	1.3932	1.6341
9	0.85	0	0.5105	1	1.4846	1.7245
11	0.90	0	0.5058	1	1.4888	1.7950
13	0.95	0	0.5049	1	1.4905	1.8553
21	0.95	0	0.5018	1	1.4961	1.9382
29	0.95	0	0.5009	1	1.4981	1.9677
exact value		0	0.5	1	1.5	2

We see that the results are converging and highly precise. Here α_0 and α_2 correspond to linear solutions, hence the values obtained by the infinite element method are exact. We can also see from Table 9 that the smaller j, the faster speed of convergence, which coincides with the result of Chapter Three.

Now let us observe the changing of K_I^* as meshes are refined. We obtain the following table.

Table 10

n	n_1	N	ξ	K_I^*
7	16	24	0.85	10.397
9	43	72	0.85	10.281
11	52	90	0.9	10.383
13	62	108	0.95	10.425
21	112	208	0.95	10.523
29	132	254	0.95	10.542

We see the results are converging too. These results were compared with the numerical results of the boundary collocation method in [12]. The boundary collocation method is another approximate method. We can corroborate both of them by this comparison. The result of the boundary collocation method for this problem is $K_I^* = 10.6$, which differs from the K_I^* of the fine mesh by only 0.6% in relative error, and from the K_I^* of even coarsest mesh by only about 2%. Those results are satisfactory.

By least square method we obtain an approximate formula,

$$K_I^* = \left[7.76 + 1.54\left(0.75 - \frac{a}{W}\right)^2\right] \sec\frac{\pi a}{2W} \sqrt{\tan\frac{\pi a}{2W}}.$$

4.5 Stokes external flow

We give an example of §1.10. Assume that the flow passes a ball, the velocity at the infinity is $u_\infty = 0, v_\infty = 1$, the radius of the ball is one and the viscosity $\mu = 1$, then the expression of the solution is

$$u = (c_1 - c_2) \sin \vartheta \cos \vartheta,$$

$$v = c_1 \cos^2 \vartheta + c_2 \sin^2 \vartheta,$$

$$p = -\frac{3}{2r^2} \cos \vartheta + p_\infty,$$

where

$$c_1 = 1 - \frac{3}{2r} + \frac{1}{2r^3},$$

$$c_2 = 1 - \frac{3}{4r} - \frac{1}{4r^3},$$

p_∞ is an arbitrary constant, and $\vartheta \in [0.\pi]$ is the angle included between the radius vector and the y-axis, cf Fig.44. The mesh is shown in Fig.45, where there are 36 triangular elements in each layer, uniform partition is taken along the circles and the mesh is symmetric with respect to the x-axis. The solution obtained by the infinite element method is also symmetric with respect to the x-axis, since the problem admits this property. In the concrete, u is anti-symmetric, v is symmetric and p is anti-symmetric. For simplicity it has no harm in assuming $p_\infty = 0$.

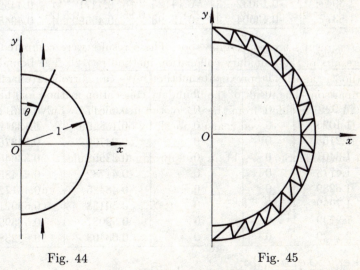

Fig. 44   Fig. 45

Take $\xi = 1.05$ and $\xi = 1.1$. The velocity is shown by contrast for $\xi = 1.05$ and for $\vartheta = 0°, 45°, 90°$, see Table 11, 12 and 13. The results are satisfactory. The numerical results for $\xi = 1.1$ are analogous, thus omitted here.

Table 11 $\vartheta = 0°$

y	u	u(exact)	v	v(exact)
1	0	0	0	0
1.025	0	0	0.00165	0.00089
1.1025	0	0	0.01258	0.01256
1.2155	0	0	0.04501	0.04436
1.3401	0	0	0.08956	0.08844
1.4775	0	0	0.14121	0.13977
1.6289	0	0	0.19643	0.19482
1.7959	0	0	0.25278	0.25107
1.9799	0	0	0.30854	0.30682
2.1829	0	0	0.36260	0.36090

Table 12 $\vartheta = 45°$

x	u	u(exact)	v	v(exact)
0.7044	0	0	0	0
0.7205	-0.01636	-0.01502	0.01794	0.01559
0.7766	-0.05901	-0.05839	0.07145	0.07004
0.8562	-0.09845	-0.09848	0.14280	0.14133
0.9440	-0.12288	-0.12329	0.21133	0.20984
1.0407	-0.13651	-0.13717	0.27631	0.27485
1.1474	-0.14252	-0.14333	0.33737	0.33596
1.2650	-0.14322	-0.14412	0.39433	0.39300
1.3947	-0.14032	-0.14125	0.44718	0.44592
1.5377	-0.13504	-0.13598	0.49599	0.49481

Table 13 $\vartheta = 90°$

x	u	u(exact)	v	v(exact)
1	0	0	0	0
1.025	0	0	0.03817	0.03614
1.1025	0	0	0.13819	0.13317
1.2155	0	0	0.24799	0.24376
1.3401	0	0	0.33998	0.33646
1.4775	0	0	0.41782	0.41485
1.6289	0	0	0.48425	0.48172
1.7959	0	0	0.54138	0.53921
1.9799	0	0	0.59087	0.58899
2.1829	0	0	0.63403	0.63238

The pressure on the solid surface is also shown by contrast, see Table 14. It is observed that the results coincide in general, but are not so precise as that of velocity. It seems because in our scheme second order interpolation is applied for velocity, but only zero order for pressure.

Table 14

ϑ	$p(\xi = 1.05)$	$p(\xi = 1.1)$	p(exact)
5°	-1.4709	-1.4470	-1.4886
15°	-1.4142	-1.3788	-1.4433
25°	-1.3228	-1.2917	-1.3543
35°	-1.1925	-1.1679	-1.2241
45°	-1.0262	-1.0100	-1.0566
55°	-0.8291	-0.8218	-0.8571
65°	-0.6073	-0.6090	-0.6315
75°	-0.3668	-0.3782	-0.3868
85°	-0.1163	-0.1369	-0.1302

The contrast of resistance acting on the ball is shown in Table 15. We observe that the results are satisfactory, and the value for $\xi = 1.1$ is even closer to the exact value.

Table 15

	$\xi = 1.05$	$\xi = 1.1$	exact
resistance	18.424	18.748	6π
relative error	2.26%	0.54%	0

In the process the combined stiffness matrix is evaluated by the iterative method. Using the iterative method of the first type with seven times of recurrence we get a result of about five significant decimal digits, then using the iterative method of the second type with three times of recurrence we get more than ten significant digits.

4.6 Square driven cavity

Fig. 46 shows a square cavity. We assume that it is full of viscous incompressible fluid. At the top the velocity u is horizontal and uniform. It has no harm in assuming $|u| = 1$ there. We assume that the flow is steady and the Reynolds number is so small that we can neglect the convection term in the equation.

Let ψ be the stream function, then it satisfies

$$\Delta^2 \psi = 0,$$

$$\psi|_{\partial \Omega_0} = 0,$$

$$\frac{\partial \psi}{\partial \nu} = \begin{cases} 1, & (x,y) \in AD, \\ 0, & (x,y) \in \partial\Omega_0 \setminus AD, \end{cases}$$

where Ω_0 is the square domain and $\partial\Omega_0$ is the boundary.

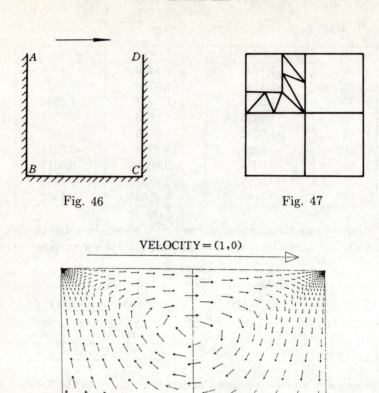

Fig. 46

Fig. 47

Fig. 48

The domain Ω_0 is divided into 4 squares. Each of them is a combined element with the center A, or B, or C, or D. The mesh is shown in Fig.47. We use the approach in §1.12 to solve this problem. The flow field is shown in Fig.48. Let us use a magnifying glass to see the details near the corners. An interesting phenomenon is discovered. Figures 49–56 are the gradual enlargement of the neighborhood of the point B. We see a sequence of vortices. Each one has an opposite direction to the previous one. We believe that we can get as many vortices near one corner as we want if we keep on doing our computation. The total number of vortices must be infinity.

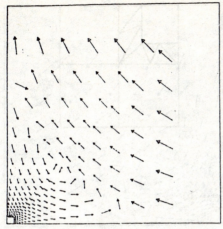

lower left quadrant of the cavity
PARTITION: 12×12; $\xi=0.83$
(LAYER: from 8 to 25)

Fig. 49

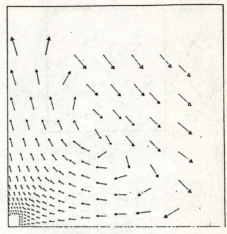

lower left quadrant of the cavity
PARTITION: 12×12; $\xi=0.83$
(LAYER: from 26 to 40)

Fig. 50

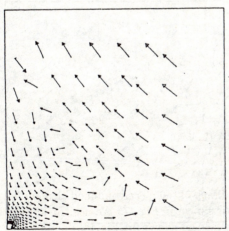

lower left quadrant of the cavity
PARTITION: 12×12; $\xi=0.83$
(LAYER: from 40 to 59)

Fig. 51

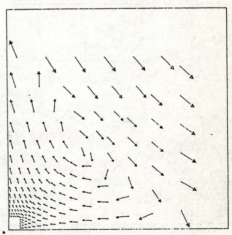

lower left quadrant of the cavity
PARTITION: 12×12; $\xi=0.83$
(LAYER: from 56 to 70)

Fig. 52

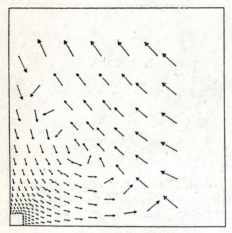

lower left quadrant of the cavity
PARTITION: 12×12; ξ=0.83
(LAYER: from 71 to 85)

Fig. 53

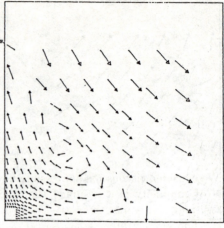

lower left quadrant of the cavity
PARTITION: 12×12; ξ=0.83
(LAYER: from 86 to 100)

Fig. 54

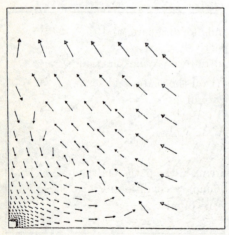

lower left quadrant of the cavity
PARTITION: 12×12; ξ=0.83
(LAYER: from 101 to 118)

Fig. 55

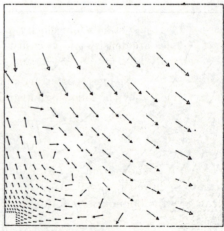

lower left quadrant of the cavity
PARTITION: 12×12; ξ=0.83
(LAYER: from 116 to 130)

Fig. 56

4.7 Navier-Stokes external flow

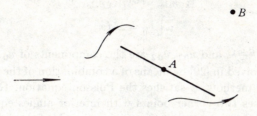

Fig. 57

We consider viscous incompressible flow passing a finite plate, cf Fig.57. For the sake of simplicity it is assumed that the flow is two dimentional. Denote by Ω_0 the exterior domain of the plate, then the fluid motion is governed by the following initial-boundary value problem of the Navier-Stokes equation [53]:

$$\frac{\partial u}{\partial t} + (u \cdot \nabla)u + \frac{1}{\rho}\nabla p = \tilde{\nu} \triangle u, \tag{4.7.1}$$

$$\nabla \cdot u = 0, \tag{4.7.2}$$

$$u\big|_{x \in \partial\Omega_0} = 0, \quad u\big|_{|x|=\infty} = u_\infty, \tag{4.7.3}$$

$$u\big|_{t=0} = u_0(x), \tag{4.7.4}$$

where $u = (u_1, u_2)$ stands for velocity, p stands for pressure, and the positive constants ρ and $\tilde{\nu}$ are density and viscosity respectively.

Let vorticity be $\omega = \frac{\partial u_2}{\partial x_1} - \frac{\partial u_1}{\partial x_2}$, and we introduce stream function ψ, such that $u = (\frac{\partial \psi}{\partial x_2}, -\frac{\partial \psi}{\partial x_1})$. Fixing an arbitrary point A on the plate, and setting $\psi(A) = 0$, we determined ψ by the following integral over curves:

$$\psi(x) = \int_{\breve{AB}} (-u_2\, dx_1 + u_1\, dx_2). \tag{4.7.5}$$

By (4.7.2) it is known that (4.7.5) is independent of the path. On the both sides of the plate we have $\psi\big|_{\partial\Omega_0} = 0$. Applying the boundary condition for u at the infinity, we can give the boundary condition for ψ at the infinity,

$$\psi(x) = -u_{2\infty}x_1 + u_{1\infty}x_2 + O(1), \tag{4.7.6}$$

where $u_{1\infty}$ and $u_{2\infty}$ are the components of u_∞. The equations with unknowns ω, ψ are

$$\frac{\partial \omega}{\partial t} + (u \cdot \nabla)\omega = \tilde{\nu} \triangle \omega,$$

$$\triangle \psi = -\omega.$$

In addition to the boundary condition (4.7.6), we have the boundary and initial conditions,

$$\psi\Big|_{x\in\partial\Omega_0} = \frac{\partial\psi}{\partial\nu}\Big|_{x\in\partial\Omega_0} = 0,$$

$$\omega\Big|_{t=0} = \omega_0(x),$$

where $\omega_0 = \frac{\partial u_{02}}{\partial x_1} - \frac{\partial u_{01}}{\partial x_2}$, and u_{01}, u_{02} are the components of u_0.

This problem is solved in [29] by means of a combination of the vortex method and the infinite element method. ψ satisfies the Poisson equation. Ω_0 is an unbounded domain and possesses two corner points with interior angles equal to 2π. Infinite and similar partition are assumed at the neighborhoods of the corner points and the infinity. Let the attack angle be 20° and the Reynolds number be 1000, the numerical simulation of the flow field is obtained. The process of vortices created incessantly from the surface of the plate and running downstream incessantly is shown from Fig.58 to Fig.64.

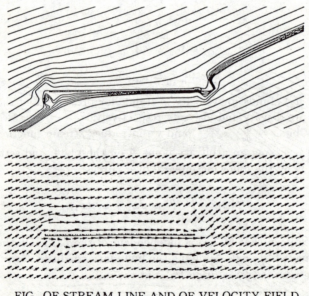

FIG. OF STREAM LINE AND OF VELOCITY FIELD
$Re=1000.0;\quad \theta=20.0;\quad t=0.00$

Fig. 58

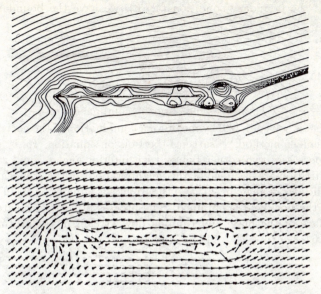

FIG. OF STREAM LINE AND OF VELOCITY FIELD
$Re=1000.0;\quad \theta=20.0;\quad t=0.20$

Fig. 59

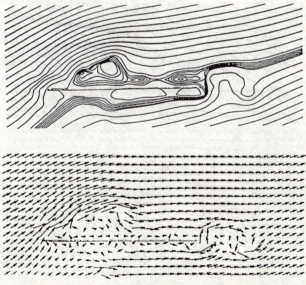

FIG. OF STREAM LINE AND OF VELOCITY FIELD
$Re=1000.0;\quad \theta=20.0;\quad t=0.40$

Fig. 60

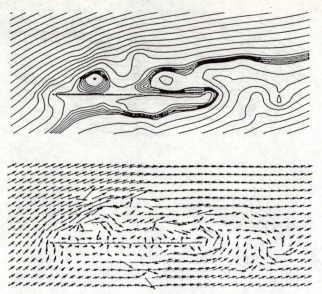

FIG. OF STREAM LINE AND OF VELOCITY FIELD
$Re=1000.0; \quad \theta=20.0; \quad t=0.60$

Fig. 61

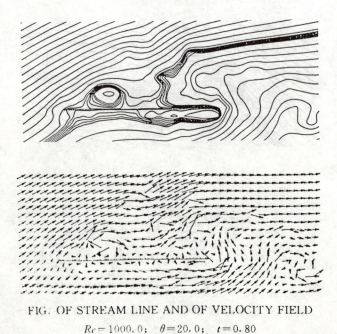

FIG. OF STREAM LINE AND OF VELOCITY FIELD
$Re=1000.0; \quad \theta=20.0; \quad t=0.80$

Fig. 62

4.7 NAVIER–STOKES EXTERNAL FLOW

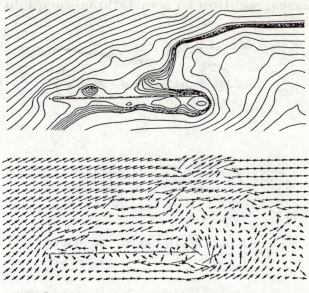

FIG. OF STREAM LINE AND OF VELOCITY FIELD
$Re=1000.0$; $\theta=20.0$; $t=1.00$

Fig. 63

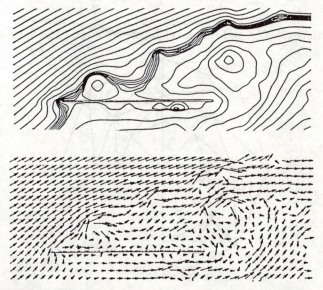

FIG. OF STREAM LINE AND OF VELOCITY FIELD
$Re=1000.0$; $\theta=20.0$; $t=1.20$

Fig. 64

4.8 A numerical solution to a variational inequality

We consider an example of §1.17. Let

$$\Omega_0 = \left\{ (r, \vartheta); 0 \leq r < 1, 0 < \vartheta < \frac{3\pi}{2} \right\},$$

$$\varphi = 0,$$

and

$$f = 16 \left(\frac{10}{3} r^{\frac{2}{3}} - \frac{64}{3} r^{\frac{8}{3}} \right) \sin \frac{2\vartheta}{3}.$$

The exact solution is

$$u = \begin{cases} 16(r^2 - \frac{1}{4})^2 r^{\frac{2}{3}} \sin \frac{2\vartheta}{3}, & (r, \vartheta) \in \Omega_1, \\ 0, & (r, \vartheta) \in \Omega_2, \end{cases}$$

where

$$\Omega_1 = \left\{ 0 \leq r < \frac{1}{2}, 0 < \vartheta < \frac{3\pi}{2} \right\},$$

$$\Omega_2 = \left\{ \frac{1}{2} < r < 1, 0 < \vartheta < \frac{3\pi}{2} \right\}.$$

Two meshes, finite element and infinite element, are given. The finite element mesh is shown in Fig.65.

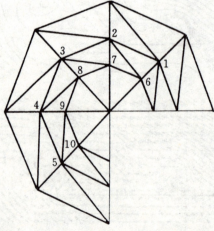

Fig. 65

We compare the approximate solution with the exact one at 10 nodes which are also shown in Fig.65. The result is the following.

Table 16

numerical solution	0	0	0	0	0
exact solution	0	0	0	0	0
numerical solution	0.1091	0.1920	0.2227	0.1920	0.1091
exact solution	0.1116	0.1933	0.2232	0.1933	0.1116

The largest relative error is about 2%.

Notes to Chapter 4

The material of this chapter is compiled from the following references: §4.1-[36], §4.2-[30], §4.3-[33], §4.4-[4],[12],[19], §4.5-[38], §4.6-[35], §4.7-[29], §4.8-[41]. [12] contains some numerical results for arched specimen which are not included in this chapter.

Bibliography

Part One   Infinite Element Methods

[1] Silvester, P., and Cermak, I.A., *Analysis of coaxial line discontinuities by boundary relaxation*, IEEE Trans. MTT, 17,8(1969), 489–495.

[2] Thatcher, R.W., *Singularities in the solution of Laplace's equation in two dimensions*, J. Inst. Math. Appl., 16(1975), 303–319.

[3] Thatcher, R.W., *The use of infinite grid refinements at singularities in the solution of Laplace's equation*, Numer. Math., 15(1976), 163–178.

[4] Ying, L.-a., *The infinite similar element method for calculating stress intensity factors*, Scientia Sinica, 21,1(1978), 19–43.

[5] Thatcher, R.W., *On the finite element method for unbounded regions*, SIAM J. Numer. Anal., 15(1978), 466–477.

[6] Ying, L.-a., *The convergence of the infinite similar element method*, Acta Math. Appl. Sinica, 2,2(1979), 149–166.

[7] Han, H.D. and Ying, L.-a., *An iterative method in the infinite element*, Math. Numer. Sinica, 1,1(1979), 91–99.

[8] Guo, Z.H., *Similar isoparametric elements*, Science Bulletin, 24,13(1979), 577–582.

[9] Wang, S.J., *"Equivalent" stiffness matrices for the neighborhood of singularity*, Acta Math. Appl. Sinica, 2,4(1979), 302–307.

[10] Ying, L.-a. and Han, H.D., *On the infinite element method for unbounded domain and nonhomogeneous problems*, Acta Math. Sinica, 23,1(1980), 118–127.

[11] Feng, K., *Differential vs integral equations and finite vs infinite elements*, Math. Numer. Sinica, 2(1980), 100–105.

[12] Ying, L.-a. and Pan, H., *Computation of K_I and compliance of archshaped specimen by the infinite similar element method*, Acta Mechania Solids Sinica, 1 (1981), 99–106.

[13] Han, H.D., *The numerical solutions of interface problems by infinite element method*, Numer. Math., 39(1982), 39–50.

[14] Han, H.D. *The error estimates for the infinite element method for eigenvalue problems*, R.A.I.R.O. Numer. Anal., 16,2(1982), 113–128.

[15] Han, H.D., *The finite element method in a family of improperly posed problems*, Math. Comp., 38,157(1982), 55–65.

[16] Ying, L.-a., *The infinite element method*, Advances Math., 11,4(1982), 269–272

[17] Shao, X.-m. and Wang, S.J., *Fourier method for the infinite element systems*, Numer. Math., A Journal of Chinese Universities, 2(1983), 160–169.

[18] Han, H.D., *The infinite element method for eigenvalue problems*, System Science and Math., 3,3(1983), 163–171.

[19] Ying, L.-a., *The infinite element method, Part I. The infinite element method for equations with constant coefficients*, Proceedings of the China-France Symposium on FEM, Science Press, Beijing, China, Gordon and Breach, Science Publishers Inc., New York, 1983, 487–541.

[20] Ying, L.a., *The infinite element method, Part II. The infinite element method of non-similar case*, ibid, 542–565.

[21] Ying, L.-a., *The convergence of infinite element method for the non-similar case*, J. Comp. Math., 1,2(1983), 130–142.

[22] Ying, L.-a. and Han H.D., *The general solutions for similar elements and the finite similar element method*, Chin. Ann. Math., 4A,5(1983), 557–570.

[23] Xu, J.-c., *Some inequalities in the Sobolev spaces and the finite and infinite element method on polygonal domains*, thesis, Department of Mathematics, Peking University,1984.

[24] Wu, S.-x., *Infinite element method for mine groundwater problems*, Hydrogeology and Engineering Geology, 82(1985), 1–4.

[25] Xu. J.-c., *The error analysis and the improved algorithms for the infinite element method*, Proceedings of the 1984 Symposium on Differential Geometry and Differential Equations, Science Press, Beijing , China, 1985, 326–331.

[26] Ying, L.-a.,*Infinite element approximation to axial symmetric Stokes flow*, J. Comp. Math., 4,2(1986), 111–120.

[27] Wu, S.-x., *Infinite element method for elliptic equations*, J. Hebei Normal Univ., 1(1986), 1–23.

[28] Wu, S.-x., *Infinite element method for parabolic equations*, Contribution to the First National Conference on the Numerical Solutions of PDE,1986.

[29] Lu, J.-q., *The construction of numerical solution of the N-S equation with singularity at corner point of the boundary*, (to appear).

[30] Liu, W.-d., *Infinite element method for the non-similar case*, thesis, Department of Mathematics, Peking University, 1988.

[31] Xu, J.-c. and Ying, L.-a., *Analysis and improvement of algorithms for the infinite element method*, Chinese J. Num. Math. and Appl., 10,1(1988), 71–83.

[32] Abdel-Messieh, Y.S. and Thatcher, R.W., *Estimating the form of some three-dimensional singularities*, Comm. Appl. Numer. Meth., 6(1990), 333–341.

[33] Ying, L.-a., *Infinite element method for elliptic problems*, Science in China (Series A), 34,12(1991), 1438–1447.

[34] Ying, L.-a., *Multigrid algorithm for the infinite element method*, Chinese J. Num. Math. and Appl., 14,3(1992), 56–68.

[35] Xie, S.-f., *Infinite element methods for calculating square cavity Stokes flow and eigenvalues of the Laplace operator on plane concave domains*, thesis, Department of Mathematics, Peking University, 1992.

[36] Ying, L.-a., *An introduction to the infinite element method*, Math. in Practice and Theory, 2(1992), 69–78.

[37] Wu, D.-b., *Infinite element multigrid algorithm for interface problem in plane*, Acta Scientiarum Naturalium Universitatis Pekinensis, 28,5(1992), 557–565.

[38] Ying, L.-a. and Wei, W.-m., *Infinite element approximation to axial symmetric Stokes flow* (II), Math. Num. Sinica, 15,2(1993), 129–142.

[39] Ying, L.-a. and Feng, H., *Infinite element method for evolution equation*, Acta Scientiarum Naturalium Universitatis Pekinensis, 29,3(1993), 257–266.

[40] Feng, H., *Infinite element method for semilinear elliptic equations on concave domains*, Acta Scientiarum Naturalium Universitatis Pekinensis, 30, 1(1994), 1–5.

[41] Feng, H. and Ying, L.-a., *Variational inequality equations with discontinuous coefficients and their infinite element solutions*, (to appear).

Part Two   Other Literature

[42] Lax, P.D. and Milgram, A.N., *Parabolic equations*, Annals of Mathematics Studies No.33, Princeton University Press, Princeton, 1954, 167–190.

[43] Gantmacher, F.R., *The Theory of Matrices*, 2 Vols, Chelsea, New York, 1959.

[44] Smirnov, V.I., *Course in Higher Mathematics*, Vol.5, Moscow, Nauka, 1959 (Russian), German edition: Lehrgang der höheren Mathematik, Berlin, Deutscher

Verlag der Wissenschaften, 1960.

[45] Varga, R.S., *Matrix Iterative Analysis*, Prentice-Hall, Englewood Cliffs, N.J., 1962.

[46] Agmon, S., Douglis, A. and Nirenberg, L., *Estimates near the boundary for solutions of elliptic partial differential equations satisfying general boundary conditions II*, Comm. Pure Appl. Math., 17,1(1964), 35–92.

[47] Wilkinson, J.H., *The Algebraic Eigenvalue Problem*, Oxford University Press, London, 1965.

[48] Stampacchia, G., *Le probleme de Dirichlet pour les équations elliptiques du second ordre à coefficients discontinus*, Ann. Inst. Fourier(Grenoble), 15(1965), 189–258.

[49] Kato, T., *Perturbation Theory for Linear Operators*, Springer-Verlag, 1966.

[50] Kondratév, V.A., *Boundary value problems for elliptic equations in domains with conical or angular points*, Trudy Moskov Mat Obsc, 1967, 209–292.

[51] Liebowitz, H.(eds), *Fracture* Vol.2, Academic Press, New York and London, 1968.

[52] Ladyzhenskaya, O.A. and Ural'tseva, N.N.,*Linear and Quasilinear Elliptic Equations*, English Transl. Academic Press, New York, 1968.

[53] Ladyzhenskaya, O.A., *The Mathematical Theory of Viscous Incompressible Flow*, English Trans., 2nd ed., Gordon and Breach, New York, 1969.

[54] Kellogg, R.B., *Singularities in interface problems*, Numerical Solution of Partial Differential Equations II, Academic Press, New York, London, 1971, 351–400.

[55] Lions, J.L. and Magenes, R.,*Nonhomogeneous Boundary Value Problems and Applications*, Springer-Verlag, 1972.

[56] Aziz,A.K.(eds), *The Mathematical Foundations of the Finite Element Method with Applications to Partial Differential Equations*, Academic Press, New York and London, 1972.

[57] Ciarlet, P.G. and Raviart, P.-A., *Maximum principle and uniform convergence for the finite element method*, Comput. Methods Appl. Mech Engng., 2(1973), 17–31.

[58] Strang, G. and Fix, G.J., *An Analysis of the Finite Element Method*, Prentice-Hall, Englewood Cliffs, N.J. 1973.

[59] Nitsche, J.A. and Schatz, A.H., *Interior estimates for Ritz-Galerkin methods*, Math. Comp., 28(1974), 937–958.

[60] Adams, R.A., *Sobolev Spaces*, New York,San Francisco,London, Academic Press, 1975.

[61] Ciarlet, P.G., *The Finite Element Method for Elliptic Problems*, North-

Holland, Amsterdam,New York,Oxford, 1978.

[62] Yosida, K., *Functional Analysis*, 5th ed., Springer-Verlag,Berlin, 1978.

[63] Girault, V. and Raviart, P.-A., *Finite Element Approximation of the Navier-Stokes Equations*, Lecture Notes in Mathematics Vol.749, Springer-Verlag, 1979.

[64] Chen, C.M. *Finite Element Method with its Analysis to Improve Precision*, Hunan Science and Technology Press, 1982(Chinese).

[65] Pazy, A., *Semigroups of Linear Operators and Applications to Partial Differential Equations*, Springer-Verlag, 1983.

[66] Bramble, J.H. and Pasciak, J.E., *New convergence estimates for multigrid algorithms*, Math. Comp., 49(1987), 311–329.

[67] Xu J.C., *Theory of multilevel methods*, Pennsylvania State University Applied Mathematics Report Series AM 48, 1989.